The Technoscientific Turn of Philosophy

Źarko Paić

The Technoscientific Turn of Philosophy

The Age of Cybernetics

palgrave
macmillan

Źarko Paić
Visual Studies
University of Zagreb
Ivanić-Grad, Croatia

ISBN 978-3-032-21080-7 ISBN 978-3-032-21081-4 (eBook)
https://doi.org/10.1007/978-3-032-21081-4

This Palgrave Macmillan imprint is published by the registered company Springer Nature Switzerland AG
The registered company address is: Gewerbestrasse 11, 6330 Cham, Switzerland

Preface

This book represents one of the key attempts at my philosophical reflections on contemporaneity, starting from the fundamental post-metaphysical concept of the technosphere as the realization of cybernetics in the process of creating artificial life. Therefore, a technoscientific turn in philosophy has been undertaken here, with its roots in the intellectual endeavours of Heidegger, Whitehead, and Deleuze. But what does the technosphere mean?

The technosphere, briefly, denotes a system of information, a recursive contingency of events, whose "essence" lies in the process of realizing technological singularity. What is the essential difference, however, between the management of history from the horizon of metaphysics as cybernetics and the navigation of post-history from the trans-classical logic of the technosphere? The technosphere should be understood as a cybernetic system whose "essence" manifests as the triad of calculation, planning, and construction. Cybernetics reaches its peak in the management of the world as a techno-poietic "human garden and non-human environment", which it controls using a system of information that enters the system of contingent events.

What is the uncanny in this concept of the technosphere, other than that metaphysics is realized in a thing that thinks differently from us, but is programmed to learn and create by itself, something we cannot do quickly or effectively. The entire problem of the technological world in which we humans exist, along with the apparatuses of non-human, lies in the infinite acceleration of the system and the environment, and the appropriate technologies of dizzying memory speed for the event of singularity. With the disappearance of humans as creators of the artistic world, a multitude of possibilities for synesthetic metamorphoses of life arise, which have already become artificial life. When philosophy is realized in the technoscientific turn, the age of cybernetics emerges.

I want to thank several friends and colleagues in the fields of contemporary philosophy, art and architecture, contemporary art theory, media studies, and technoscience, with whom I had intensive discussions on these issues, which are not only theoretical but also existential. I have found many of their insights to be precious stimuli in my own reflections, and this experience represents a happy moment of genuine dialogue with the very matter of thought that appears in various aspects of the relationship between philosophy, science, and art. I owe a special debt of gratitude to my executive editor, Amy Invernizzi, at Palgrave Macmillan for her care and helpfulness in publishing this book.

Zagreb, Croatia
February 2026

Žarko Paić

Contents

CHAPTER 1

Introduction: Theses on the Technosphere as the End of the "Cosmic Epoch"

1.1

The technosphere transcends thought as a reflection "about" Being and "on" being, because it is neither an absolute system nor a whole of scattered, fractal, and fragmented elements. Since it denotes at the same time neither "technique" (an instrument for some other purpose) nor "technology" (applied science in practice as the real state of the world), then its "essence" lies in the realization of metaphysics as "immanent transcendence" in the idea of autopoiesis. With the technosphere, the "age of the world" (J. W. Schelling) and the "age of the primaeval earth" (E. Husserl) cease. Its "age" becomes that which belongs to the end of the "cosmic epoch" (A. N. Whitehead).

1.2

The entire so-called history of human adventure in space, along with the technosphere, represents a transition to a posthuman condition that transcends the boundaries of living and the dead. The only remaining difference becomes that between artificiality and life, and "life" ceases to be a God-given and nature-endowed event of the power to unfold that which has the characteristics of the "archetype of the unconscious" (C. G. Jung). Artificial "life" arises from the work of an "artificial brain" that

Ž. Paić, *The Technoscientific Turn of Philosophy*,
https://doi.org/10.1007/978-3-032-21081-4_1

includes rational and intuitive thinking. There can no longer be a distinction between *a priori* and *a posteriori* because in the age of "immanent transcendence" there is nothing outside the "logic of processuality".

1.3

Everything is becoming (*devenir-Werden*) except for the becoming itself that occurs from the complete openness of the event beyond the difference between *Ereignis* and *événement* (M. Heidegger vs. G. Deleuze). Therefore, philosophy at the end of the "cosmic epoch" is nothing other than the tautology of the quadrilateral arising from the cybernetic principles of action: information, feedback, control, and communication. This tautology also encompasses a synthesis of event and autopoiesis. Moreover, the event produces events, just as the self-production of life becomes the production of self-realization, the idea becoming a practice in the technoscientific construction of worlds.

1.4

When there are no more speculative-reflexive or dialectical differences and oppositions between universality, particularity, and individuality, only the becoming-process-creation of intensities and levels (plateaus in the sense of Deleuze's and Guattari's "image of thought") is at work, no longer of the one-and-the-same, but of singularity-as-difference. The question of the "meaning of Being" and the problem of the human-too-human are over, because all anthropologies, from structural to philosophical and cybernetic, are not only outdated conceptual tools, but also a colossal failure to save the so-called Human from his aspiration for the inhuman. The futurist imperative of the poet Fernando Pessoa must be taken very seriously: *I want to be a machine!*

1.5

Why is there so much tautological "enthusiasm" and "indifference" in understanding the technosphere as a triad of calculation, planning, and the construction of artificial life? Because the event of autopoiesis has no resistance in the reality determined by the theories of deterministic chaos and entropy, which is best illustrated by Nietzsche's famous parable of how "the desert grows". The resistance does not arise from thinking as an event of openness of the conception (*Besinnung* in Heidegger) of the world, because everything that such this-or-that, object-physical, or post-metaphysical thinking (from Parmenides to Severino) can show as an alternative to the technosphere as the thinking of "rational intuition" boils down to the mysticism of the irreducibility of spontaneous freedom as an event of indeterminacy and contingency and therefore falls under the auspices of what Heidegger in a lecture in the 1950s called releasement (*Gelassenheit*).

1.6

The technosphere encompasses absolute and total Power, and not any pseudo-complexes of social articulation of human will that man, as a subject-substance, elevates to the rule of the non-human. Human, therefore, necessarily transcends the limits of his own "humanism". It is thus no coincidence that the late Marx, after his *Grundrisse: Foundations of the Critique of Political Economy* and *Capital*, as well as the late Heidegger in his *Letter on Humanism*, clearly saw the consequences of the turn in the very essence of metaphysics. This thought unites what Vanja Sutlić calls the practice of work as scientific history with the idea of the openness of events. Both Marx and Heidegger abandon the philosophical discourse of metaphysics, the former by explaining this world in the language of the science of political economy but as a critique of its essence starting from the concept of practice and reality, and the latter by claiming that philosophy has ended in the idea of cybernetics and that all that remains is reduced to the openness of thought of the "second beginning" (*andere Anfang*).

1.7

The technosphere, therefore, cannot be neither a machine nor a apparatus, so it seems entirely unnecessary to insist on linguistic differences through Greek-Latin-Slavic variants that would be most appropriate for explaining the logic of the operation of this "spiritual machine" that thinks of itself by calculating-planning-constructing states, and does not "imitate" and does not "represent" but always what is given and gifted as God's or the work of nature in the sense of thinking as a "discovery" of a new creation "with a first beginning". Furthermore, the technosphere cannot be merely a copy of a computer, and its computational logic applies to the outside world. Instead of language as an "organ" for conceiving existence through myth, religion, art, and philosophy, the technosphere rests on the pragmatics of pictorial thinking, driven by the "mystery of facticity" or the power of contingency—the concept of code.

Without coding, life as a set of genes and information would not be possible, and action in the technical world would not be possible. That's why the technosphere emerges, creating new relationships in the circuits of life, starting from the embeddedness and disembodying of fractal fragments, not parts. Everything becomes simultaneously open and closed, and only through coded reality can the system and environment of this biocybernetic network of relationships be set in motion. Instead of a totality that vertically determines the constituent parts of its "media empire", there is a network that constantly expands horizontally into infinity like a "fractal republic".

1.8

The "philosophical" concept of the technosphere derives from the technical conception of information as pragmatic instruction for action. Information is therefore information. This tautology corresponds to the theory of reversal against the image, also known as the *Iconic turn* (science of the image, *Bildwissenschaft*), which marks a breakthrough in the understanding of images. Information is information in its constant production of the "new" because just as the cosmos determines the breakdown of elementary particles, so the entire network of social relations in the technosphere system is marked by the networking of fractals of information in the transversally placed structures of life itself. Thinking as a phenomenological noesis cannot be directed at external or internal objectivity as a phenomenological noema. Instead, we encounter only two fundamental concepts in the information-feedback-control-communication process: quantification and visualization.

Everything becomes, as Max Planck claimed, a question of number, and everything is, as digital constructivists claim, a question of image. However, the (digital) image is neither mimesis nor representation, but an information product of the synthesis of calculation-planning-construction: 0–1. That is why all images today, as media theorist Friedrich Kittler shows, are calculated images. However, this "calculation" arises from a deeper, "ontological" matrix of thinking-as-calculation (*Rechnen*, as Heidegger puts it in the book *What is Called Thinking*?).

1.9

Thinking as calculating cannot be just one possibility of thinking at the end of the "cosmic epoch" of the world, and it is the one that, in technoscience, rules over all areas of thought in general, the second possibility being the one to which philosophy and art belong (*Dichten*, says Heidegger). It is a new "absolute" that, unlike Hegel's absolute, does not equate the concept of wisdom (*sophia*) with knowledge (*episteme*). Instead, it *autopoietically* creates an artificial reality, beginning with new thinking as the third order of cybernetics (*tertium datur*). The opinion of thought is represented in the traditional metaphysics of the permanence of being. Thought, on the other hand, as a thinking construction of the technosphere, assembles the collapsible and unfolds the collapsible into a new order that Gilbert Simondon, the most crucial philosopher of cybernetics and predecessor of technosphere thought, calls metastable equilibrium.

It consists of contingencies, singularities, chaos, and entropy. Nothing beyond that can be said of "existence". After all, existence no longer has any human-to-human characteristics, because the technosphere "is" a triad of cosmic-bio-techno-evolution: animal-man-machine. Cordyceps, robots, and Overman also exist. The immanent nature of information refers to its exchange and multiplication, akin to a virus or an Alien. What is true of cordyceps—parasitic fungi that multiply like a plague in the very bodies of insects and thus destroy their colonies seems analogous to what happens with informational or corporate capitalism today.

1.10

The time of the technosphere is no longer reducible to the ecstatic-dimensional triad of past, present, and future, which for Human as a thinking existence always signifies "my" taking on being, starting from the temporality of the coming. The technosphere denotes a combination of the Greek words/concepts *téchne* and *sphaira*. It refers to the circular movement of the planets in their "contingent necessity", which resembles, both externally and internally, the Nietzschean concept of eternal recurrence. However, the technosphere no longer presupposes a distinction between time and space, but rather their synthesis into space-time, in the sense of cosmology and particle physics. Hence, the technosphere does not always imitate or represent what is unchanging and constant.

The technosphere profoundly alters our understanding of both space and time. An example that becomes paradigmatic in the modern age is Leibniz's conception of infinite space as a dynamic quantity, in contrast to Newton's conception, which is more akin to a container for masses and objects. Hence, time is determined by the concept of an event as a process of becoming new, transcending the metaphysically understood eternity. Everything accelerates and implodes in the very core of becoming, from stars to black holes, from empires to republics, from Human as *homo faber* and *homo ludens* to *homo kybernetes*. The space-time of the technosphere becomes stretchable and indeterminable. In its hyper-plasticity, the image of thought becomes that of a "scanning" of life, constantly transforming as planets metamorphose. There is no possibility of changing this except through the radical transformation of the technosphere itself, which thinks its thought within the conceptual-categorical system and the environment of Nothing.

1.11

The technosphere represents a fundamental aspect of the third order of nihilism, characterized by a collapsing-and-creating event that, like Plato's idea, embodies the very indestructibility of thought itself as the essence of life in the universe. Nothing has its meaning without thought as the first and last reason for the existence of life itself, even when, instead of divinely understood nature and mind, we assert, like Jean Rostand, that all this is here and there only a "ridiculous adventure of protoplasm". Thought becomes an unprecedented case in the history of Being as an event, and all prospects for the coming future of what we still call philosophy, as the conception and as the existential thrust of freedom, depend on it.

Throughout nine chapters of the book entitled *The Technoscientific Turn of Philosophy: The Age of Cybernetics*, these basic settings are developed:

Technosphere can no longer be thought of in terms of the binary oppositions of classical metaphysics, including "Being", "substance", "linearity", "reciprocity", and "symmetry". Simondon was the first to point out that, instead of the steadfastness of Being, he began to point to a metastable balance in the essence of things. Instead of onto-genesis, he advocated the techno-genetic structure of objects in a series of events. Everything valid within the modern reproductive matrix of technology, based on the idea of infinite multiplication of the original, is now shifted to processes of creating "a lifetime". Metaphysical thinking undergoes radical changes from the beginning of the new Era, so that the cogito, the subject, and the absolute become a condition for the possibility of thinking «about» Being-God-World-Human. The axiomatic power of the metaphysics of subjectivity is developed based on the scientific-technological structure of consciousness. It begins with the concept of certainty, which must be mathematically and physically proven to harbour no doubts regarding its ontological status.

Therefore, instead of reproducing the Being and replicating things, a new way of cybernetic management of history has been established, without the purpose or the pre-existing plan of God's creation. The key process linking information and the genetic code is *cloning*. It must be a creative copying process that provides an opportunity to perfect the species, driven by the subconscious, as it merges the biological nucleus and the technical structure of the genetic stem cell (DNA). In a series of new forms and discourses of contemporary art, some explicitly referred to a new constellation as a genocentric and cognitive adaptation to the idea and practice of life. Among them, it will be necessary to distinguish between post-genomics and transgenic art. Aesthetics should be transformed into aestheticism, and thinking should be reconfigured as cognitive protocols for predicting events based on algorithmic accounts (computations). That turn introduces the openness of the new corporeal experience. We can say, however, that the concepts inherited from phenomenology and psychoanalysis are no longer usable. The body in the aesthetic mode of production of thought goes beyond the intentionality of objects and the traumas of the Real.

When the world is confronted with immeasurable attacks of violence, cruelty, and destruction of life that have their foundation in the mental order of ordinary Being, the only answer should be to try to overcome the source of the very foundation of what constitutes evil. But what if the answer to the phenomenon of evil in the modern world is impossible from any traditional metaphysical perspective that has emerged throughout Western history, by the thought of theodicy? In analogy with Heidegger's thought of the end of metaphysics as onto-theology, Lévinas speaks of the end of theodicy. It is the outcome of just such a history of salvation. It finds its meaning in the idea of God's providence/plan, another name for *logos*, mind, spirit. The deconstruction of such a way of establishing history means the openness of a situation which cannot be only radically different from the foundations of the West in general, but also distinct from the "different" as Same.

CHAPTER 2

Homo Kybernetes from onto-Genesis to the Posthuman Condition

2.1 Metaphysics and Tehnosphere

Without thoughtful reflection on the challenge posed by the relationship between technology and (post)humanity, it seems impossible to maintain a relationship between technoscience and the humanities that would not, in one way or another, become non-contemporary. Although the prefix post is in parentheses and humankind is without it, it seems evident that, today, there is still something that does not reduce human prerogatives and achievements to the logic of technological singularity. As if this play between what comes after the human and what is here and now human-to-human, to use Nietzsche's famous statement, points to the necessary difference between what constitutes the essence of posthumanism and humanity at all, which becomes the highest reach of technosphere rule in managing societies and culture with the help of artificial intelligence. But what if this slight difference no longer exists, or is suspended and neutralized? What, furthermore, if a Human has never been able to separate in historical development, i.e., what if *we have never been humane*, as Donna Haraway claims, or what if *we have always been cyborgs*, as Stefan Lorenz Sorgner puts it? (Haraway, 1990; Sorgner, 2022).

While there is a relationship between the human and the non-human, the metaphysical structure of thought is at work, even if Being is replaced with the concept of information and beings are understood as a multitude of pure contingencies. A relationship always denotes the possibility

Ž. Paić, *The Technoscientific Turn of Philosophy*,
https://doi.org/10.1007/978-3-032-21081-4_2

of a union at a distance. Synthesis, therefore, presupposes a real state of analytical separation, for example, of different substances in some hybrid assemblies such as contemporary technoscience with biogenetics, biocybernetics, and biosemiotics. When discussing the relationship between humans and non-humans, we must begin by assuming that humans serve as the model for understanding the latter. Human, in so-called human nature or its essence, cannot be understood within metaphysics in any other way than essentialist. This means that his position within Being and his being are always given to him spiritually and biologically, i.e., vertically, in accordance with Aristotle's famous definition of *animal rationale*. Furthermore, the relationship between the two, beginning with human primacy, is not only horizontally reduced to biologism, from which psychologism and anthropology follow. Heidegger attempted to eliminate this within the framework of the systematic "destruction of traditional ontology" in *Being and Time* (*Sein und Zeit*) (Heidegger, 2018).

The reason should be seen in the fact that he considered that, by doing so, "Human" or being there (*Dasein*) fails in his thinking and thereby reduces to precisely what he wanted to overcome, namely Aristotle's definition of *animal rationale*. Biologicism denotes the reduction of Human to naturalism, that is, to the biological factors of the whole of his own; psychologism becomes a reduction to the unconscious articulation of language, which reduces Human to a pre-and-post-rational living being, while anthropology means the decrease of Human to the subjectivity of the subject by which all other beings appear at a lower level of cosmic-biological evolution and thus make him the master of nature as a whole. Heidegger's critique of the reduction of the meaning of Being to the so-called life processes in their inclusive vitality affects the triad of biology, psychology, and anthropology, as they regard a Human as insufficiently and inauthentically understood by this conceptual-categorical "rationalization" of language. Any reduction of the Human in man to the non-human necessarily fragments his integrity and irreducibility as a being among beings. However, every reduction cannot be the same as the annulment of human authentic existence in the openness of freedom. To reach the purity of the human essence, phenomenologically speaking, it is necessary to understand that this essence originates in the dark core of the non-human, a concept already recognized by pre-Socratic thinkers in their discussions of the relationship between demons and humanity. We should remember that Socrates was also constantly urged by his demon to express his thoughts freely in public.

We shall see that the question of the relationship between humans and non-humans cannot be resolved by a vulgar distinction between the biological and the post-biological in the context of the relationship between humans and post (humanity). The problem with this relationship lies in the distinction between the force of life (*physis*) and the idea of life (*bios*), from which both relata derive. Humans, in the metaphysical sense, continually rise above the limits of biologically understood life, transcending mere survival at the animal level. Therefore, it seems self-evident that from this, every form of animalism will be considered as a loss of so-called human dignity. This term has its origin in connection with the ethical-aesthetic doctrines of the Renaissance notion of humanism in Marsilio Ficino and Pico della Mirandola because dignity is associated with the high rank of a human being against the bare life of an animal and the brutality with which the meaningful life of Human in the historical-epoch sense is threatened by the fall to the level of bestiality, refers to the good and the beautiful that transcends the limits of life. Therefore, ″human dignity″ represents an ethical-aesthetic ideal of living. This transcends the limits set by biological affects and drives. After all, in borderline situations when a person defies tyrannical forms of government with their own life, when they are thrown into shackles or slavishly humiliated, the last thing that remains is a person's human dignity as an existential experience of indomitable freedom. It seems evident from this that life in its complex contingency has no meaning without the originally conceived freedom. However, this does not mean that the biological organization of human existence is considered null and void for man and humanity in general (Paić, 2022b, pp. 1–36).

The distinction between human and non-human denotes what unites spirit and life, understood in a metaphysical sense, even in the realization of metaphysics in cybernetics. Namely, the spiritual always refers to the "rational", and the vital to the "animal", if we follow the logic of Aristotle's definition of Human. This problem of the relationship between the cognitive possibilities of Human's existence and the biological-physical limits of his physicality should by no means be lightly declared a relic of Cartesian metaphysics. This assemblage has been established in that way as a system of signifiers with which the concept of self-consciousness presupposes the necessary predominance of *res cogitans* over *res extensa*.

In contemporary neuroscience and neurophilosophy, this dualism is critically overcome by referring to the complex structure of interaction between thought and its natural environment, starting from the notion of a network that has neither a centre nor an edge. However, despite this "new" anti-essentialism that aims to abolish the distinction between human and non-human by requiring a different notion of all post-metaphysical concepts, such as network, interactor, communication, and cybernetic corporeality, we still live in two worlds. These worlds intersect and permeate one another. For this reason, it seems that the most crucial philosopher of cybernetics, Gilbert Simondon, was right when he considered that the future of the relationship between humans and thinking machines does not mean the abolition of humanity in the dominance of superior technology, but an attempt to build coexistence as the only meaningful way of coherence of the analogue and digital world.[1]

If, therefore, we try to understand the non-human by starting with the human, this is self-explanatory. After all, Heidegger, in *Being and Time*, attempted to demonstrate that the meaning of Being within the horizon of original temporality can be explained only if the essence of being there (*Dasein*) is first disclosed in its existential-ontological spheres, including project, care, and openness to the coming future. What is different, alien, and foreign to the human being cannot be thought about from its own logic, *par lui-même*, but only in a roundabout way. The non-human is therefore no longer merely a negation of man's ability to spiritually and biologically govern his existence, system, and environment of beings. There is something uncanny (*Unheimlich*) in that matter.

Namely, suppose a Human cannot be the result of any biological-animal intervention and reduction of his essence to the non-human in the sense of things and objects. In that case, his ontological-temporal openness encompasses an entirely different condition, both from the reduction to God in the sense of pure spirituality and incorporeality, and from the diminishment to technology in the sense of pure applicative objectivity arising from the subject-substance of modern sciences. God and technology are metaphysical ways of understanding Being as *téchne* and *poiesis*, from Plato to the end of metaphysics in cybernetics (Severino, 2016, pp. 208–209).

When we have this in mind, we can see that the problem with understanding of the non-human is primarily that what is in the *posthuman condition* can no longer be measured by the parameters in which humans traditionally think and express themselves metaphysically. What are these parameters? Undoubtedly, the autonomy of the human essence was still, for modern speculative thought from Kant to Hegel, a question of the sustainability of anthropology in a pragmatic way. Humans as subjects of knowledge of the external world were considered a substantial set of reflective practices, and even for Blaise Pascal, human dignity consisted solely of rational thinking. In other words, rationality becomes a condition for the possibility of human existence in the technical world, which, at the same time, means that science and technology are the key parameters of human potential for progress and development in the historical movement towards reaching *object X*.

The paradox is, therefore, that already with the entry into the scientific-technological age, the human, i.e., humane, became an artificial construction, and not a natural act of revealing the secret of Being. Therefore, the *posthuman condition* cannot be merely the realization of Western metaphysics' potential, as completed in cybernetics. What distinguishes the non-human from the human is the absence of non-human qualities, such as animalism and the bestiality of a creature or a monstrous being, which embodies evil in a sublime state of chaos and entropy. On the contrary, the *posthuman* denotes a becoming (*Werden, devenir*) that transcends any distinction between living and non-living, nature and culture, and Being and beings. It is therefore not at all accidental that in posthumanism, instead of God in the metaphysical sense of the first substance understood, the creator of the world and Human in it, we speak of *singularity* as the Omega point of joining the onto-theo-cosmo-anthropological perspective in the form of an absolute *technosphere* (Paić, 2022a).

What does it mean? *Singularity* assumes that the universe's becoming, as evidenced by astrophysics and cosmology, is a unique and irreversible process of the creation and destruction of matter and energy. Theories about black holes, therefore, represent a cognitive-theoretical path to understanding how supernova explosions trigger the implosion of dark matter. From a philosophical point of view, we encounter a precise movement of the very thing of thought from the speculative-reflexive analysis of the progress and development of the primordial and its form as *eidos* and *morphé* to the last station in life, regardless of whether it is about stars, plants, animals, or humans and their self-consciousness. The second law of thermodynamics, along with the concept of entropy, reveals precisely what Nietzsche metaphorically referred to as the "desert grows" statement. This means that, in its stable state, the order inevitably collapses and, as Simondon claims, can be sustained only in a *metastable equilibrium*. Simondon always thinks starting from (1) transduction (of Being); (2) trans-individuation (of beings), and (3) metastability (of information) in the transformation of the order of metaphysics (Simondon, 2005).

The fundamental concept by which singularity can be measured in the cosmic-human perspective is the basic parameter of the *posthuman condition*. Of course, this should be the term with which cybernetics begins and ends, while at the same time, the entire original structure of Western metaphysical thinking collapses with it. It is about the concept of *information* as an absolute measure of all measures in the age of the technosphere. Without information, there is no feedback loop in the life of homeostatic systems, no possibility of external or internal control over the process of event creation, and, finally, no possibility of interactive communication. It is therefore no coincidence that Norbert Wiener, the founder of cybernetics, stated that the key, third concept for the self-rule and self-management of systems and the environment, following matter and energy, is precisely the concept of information. Hence, informational entropy has become a categorical framework for the path to absolute singularity, in which the cosmo-biological merges with the anthropo-technology (Rieger, 2003).

If, therefore, Aristotle's definition of Human as an *animal rationale* placed at the centre of the entire thinking of the West from the ancient Greeks to the beginnings of cybernetics in the relationship between the concept of a living being or an animal that thinks with what makes such thinking operative and pragmatic, then this connection between the spiritual and the material, or biological has become in the age of artificial intelligence not only obsolete, but also fundamentally inappropriate for the explanation of the essence of that which is no longer non-human, nor can it be determined by something that only exceeds the limits of the Human, as established by the concept of *transhumanism* (Sorgner, 2020). Why? Precisely because a "Human" can no longer be at the centre of the world, as it was in the modern era, in which he appeared as the subject of thought at the origin of all possible and real paradigms for understanding the relationship between Being, God, and the world. Today's discourse on the concept of the end of the *anthropocene* constitutes credible evidence of this (Stiegler, 2018). The loss of the centre of Human goes hand in hand with the gain of his so-called substitute being, that is, becoming something that can no longer be merely human as opposed to a non-human.

When the distinction between Human and machine is gradually lost, we encounter the hybrid status of all beings in the *posthuman condition* of technological worldliness. The *animal rationale* can no longer be a condition for the possibility of all kinds of anthropology, from philosophical and structural to cultural and cybernetic. In its place has come a concept that no longer elevates man to the divine heights achieved by Renaissance humanism, from which today's humanities originated. After the period of establishing metaphysics, as announced by Nietzsche with the term "overman as the meaning of the Earth" (Nietzsche, 1999, p. 14), the dizzying realization of the project of becoming an interplanetary nomad began, one that rests on the dynamic notion of infinite speed, rather than static inertia (Deleuze & Guattari, 2005, p. 198).

To be able to fly and transform our physical appearance, we must first become Other, as unconditionally different from the historical-epochal worldview. This also means that Being is replaced by the process of infinite becoming. *Instead of ontogenesis, techno-genesis is at work.* "Human" no longer "is" something in the sense of intentional consciousness, which, with Husserl, distinguishes between *noesis* and *noema*, consciousness and the object of consciousness. His "fate" now lies in the contingency and emergence of life itself as a construction of events. If, by contrast with Simondon, we argue that a robot does not exist, then technological uploading transforms humans into information producers and users of an operating system. In this system, information circulates and becomes pragmatic knowledge only within specific contexts.

His "case" might always be possible to think of pure singularity and individuation. Therefore, the term *homo kybernetes*, which corresponds to this "case" of a complete transformation of the ontological place of the human in post-history, cannot be determined from the linear development of a robot controlled remotely by artificial intelligence. On the contrary, this "Human", in his cognitive self-organization and self-management, becomes an autonomous synthesizer of the logic of the technosphere.

Beyond all metaphysical definitions, its fundamental ontological feature is that it integrates information and communication into a global, planetary code of networked life. In the tendency-latency, *homo kybernetes,* therefore, becomes a visualized event of synthetic thinking, as the production of virtual worlds. That means the final farewell to every dichotomy: human-non-human, living-artificial, and natural-cultural. At the end of history, as a metaphysically determined journey from the first cause to the last purpose (*arché-eschaton*), there remains only the endless wandering of "Human" who, instead of the secret of language, is mastered by the absolute visualization of life from its origin to its disappearance. The image now precedes language as the *technosphere* precedes the biosphere, becoming its only remaining pledge of redemption and deliverance from ruin.

2.2 Stefan Lorenz Sorgner on Transhumanism

The emergence of the *homo kybernetes*, with which both posthumanism and transhumanism engage in terms of kinship and differences, indeed cannot be understood without the following essential assumptions. I conducted my analysis of this problem explicitly across five volumes *of Technosphere* (Paić, 2018–2019). For this consideration, it will be sufficient to indicate the following interrelated main ideas. After all, throughout the history of human thought within the circle of Western metaphysics, we encounter, as Ernst Cassirer would say, five fundamental symbolic forms in which the issue of the meaning of Being and time is crystallized. These are myths, religion, art, philosophy, and science. Each of these matrixes of thinking is based on correlative concepts—myth based on primordial tales (*Sages*), a religion on faith in otherworldly gods and God, art on the representation and presentation of the world in images and the language of the senses, philosophy is based on the conceptual articulation of Being, beings and essence of man, and science on the idea of knowledge as a path towards reaching the absolute truth about *object X*. This quintet of historical-epoch thinking corresponds to the number of senses in Human. It is not at all surprising that the sense of sight will become key to what the technosphere signifies: the visualization of events in synthesizing the world as *a priori* and *a posteriori* form of thought that includes calculation, planning, and construction.

However, what would now be decisive for moving beyond metaphysics to the other shore is that the technosphere represents the possibility, reality, and necessity of the cybernetic fourth: information, feedback, control, and communication. This suspends and neutralizes the four causes in traditional Aristotelian ontology, namely the *causa formalis, causa materialis, causa efficiens, and causa finalis*. At the same time, the technosphere, as a "sixth sense" or a post-metaphysical symbolic form of thought, transcends the entire history of metaphysics in all its aspects. Hence, it is its realization in the sense of the dialectical notion of the abolition of opposites and contradictions (*Aufhebung*) in the synthesis of the triad (*tollere, conservare, elevare*). Still, it is even more about the emergence of a new process of thinking, immanent in the irreversibility of the movement of thinking itself. In this way, paradoxically, Hegel, Heidegger, Wittgenstein, and Deleuze are reconciled, because what transcends any form of rootedness in space and time and thoughts intuitively starting from the rational core of artificial intelligence, can only be "founded" on *visual pragmatics knowledge*, and not in language as a telling of Being. Therefore, *homo kybernetes* can no longer be a "Human" or a "being there" (*Dasein*), nor can his post-humanity be understood solely by beginning with the transhuman turn in the essence of the cosmic-technological evolution of nature itself as the source of all kinds of cognitive-sensitive transformations.

To show how the logic of *homo kybernetes* actions unfolds in the process of this "new" thinking that starts from the premise that we have always been cyborgs, we will briefly look at the analysis of Stefan Lorenz Sorgner, the leading philosopher of transhumanism today. As a reliable witness of this way of thinking, all the more so because he has behind him a study of Nietzsche and the concept of the overman and because in a creatively stimulating way he combines the analytical and continental philosophy, Sorgner has in mind that the idea of humanism became not only inoperable already in the twentieth century after all the interventions of Heidegger and Sloterdijk, but primarily because the emergence of the emergent digital technologies contributed to the grounding of metaphysics and the dismantling of speculative metaphysics with Hegel as the pinnacle of that direction of thought.

What Sorgner calls "philosophical problems" in the discussion of transhumanism refers to the ontological, cognitive-theoretical, and practical-production aspects of the change in the very essence of human substance and spirit. In the book with the programmatic title, *We Have Always Been Cyborgs: Digital Data, Gene Technologies, and an Ethics of Transhumanism*, we can discern how the relationship between cybernetics, biogenetics, and bioethics is attempted to be considered, conditionally speaking, since the categorical-conceptual order is established with the formal primacy of information. It should be clear that this is not about the Kantian way of demonstrating the triad of thinking, but about the necessary convergence between ontology, epistemology, and ethical-political postulates. How, then, does Sorgner determine what the essence of transhumanism is concerning metaphysics in general and especially relating to different positions on the same "philosophical problem"?

Interestingly, Sorgner points out that, since Aristotle's *Nicomachean Ethics*, the problem we face with transhumanism cannot be the utopia of immortality, but rather the pursuit of such a life with improved health in both physical and cognitive senses.

> Yet neither cryonics nor mind uploading appears to be the most promising approach for achieving this goal. Given this, immortality is not a viable option. We cannot even conceptualise the notion of immortality within a naturalistic understanding of the world. Immortality implies that either humans cannot or must not die. Both options are absurd when considered from a naturalistic perspective. In this case, the universe began with the Big Bang. It has expanded since then, and eventually the expansion will slow, allowing the entire universe to come to a standstill. A different theory suggests that at a certain stage, the expansion process will transition into contraction, ultimately leading to a cosmological singularity—a point of immense density. In both cases, it does not seem that there is a realistic option for human beings to survive. (...) Hence, it is clear that immortality in its literal sense cannot be a realistic option for most transhumanists. (Sorgner, 2022, pp. 6–7)

Immortality, therefore, is excluded as the realistic solution. However, it should be pretty evident that the aspiration to reach the perfect state of a good life, led healthily and effectively, cannot stop at the virtues of prudence (*phronesis*) and perhaps even stoic *ataraxia* in the face of the inevitable fate of being chained to nature and the biological constitution of the human species. Moreover, this version of transhumanism advocated by Sorgner tends to be both "realistic" and "anti-naturalistic". At the same time, it cannot be denied that the concept of information in cybernetics is a result of digital constructivism. What is offered in the utopian horizon of posthumanism, on the other hand, for example with the postulate of singularity in the sense of the transition of the human to the *posthuman condition* by transferring the individuation of man as a person into the operative program of cyborgization and computer simulation, is not evident to Sorgner the right measure between ontology, cognitive theory, and ethical-political actions. Therefore, it seems obvious that the utopian dimension of transhumanism, which posits immortality, is rejected. Instead of this impossible goal, which, from the point of view of the natural-cosmic evolution of the universe, is even unattainable and not just absurd, Sorgner defends.

> A non-utopian version of transhumanism, as I regard utopias as extremely dangerous. Therefore, the present is sacrificed for a future that is unlikely to be realised. (Sorgner, 2022, p. 7)

If, however, we view transhumanism as a philosophical consideration of the possibility of crossing the boundaries of the human-too-human, then it will not be surprising at all that the only two techniques in the utopian horizon of reaching post-humanity are related to the notions of immortality or the singularity of humanity as a species. Namely, both cryonics and uploading of the human mind are technological postulates, not actualized means of technological development at the current level of knowledge about the impossible. Moreover, these techniques appear mutually convergent insofar as they assume the revival of a dead body and the transfer of its soul or spirit within the logic of *homo kybernetes*. When we talk about the primordial notion of man, then we have in mind a mythical-religious image of the world in which this being is in mediation with other beings with the help of that natural-biological and technological approach to the supernatural, i.e., the divine, like the mythical Icarus as the prototype of human flight into space or like a centaur as a mythical creature composed of two substances.

Transhumanism should be understood, to that extent, as the logic of the transgression of metaphysics at its end. What these opinions cross is the boundary between the animal-human body and the human soul-spirit, creating a hybrid complex that is no longer human in the traditional metaphysical sense, yet is not even a posthuman being, like a cyborgized creature from an SF dystopia. It becomes, therefore, a transitional state between two mutually close and distant worlds. For Sorgner, the analogue age of the *Anthropocene* and the digital age of the technosphere correspond to the Carbon Age and the Silicon Age, respectively (Sorgner, 2022, pp. 22–109). Of course, these are metaphors that point to a distinction in the understanding of transhumanism, starting from what belongs to the technological postulate and what belongs to bioethical judgements about the limits to which the imaginable installation of implants in the body can go, as well as surgical-pharmacological interventions to improve the structure of the human body.

However, everything that has been said remains insufficient for us to initiate a proper philosophical discussion of the essence of transhumanism in the ontological-epistemological sense. Why? There is no answer to the question of why we have always been cyborgs. Sorgner claims that

> Our transformation into cyborgs has been ongoing since the emergence of *Homo sapiens.* Acquiring a language is the first human upgrade. (...) The term 'cyborg' means cybernetic organism. The word *"cyber" originates from the ancient Greek kybernētēs, meaning "helmsman" or "pilot."* So a cyborg is a governed, steered organism." (Sorgner, 2022, p. 9)

The governed simultaneously governs by interacting with other interactors in the network of life processes. The cybernetic way of thinking always exists in dualities, rather than in trinities, as in previous metaphysics. The logic of the digital world, after all, rests on the binary code 0–1. Cyborgization, therefore, means programming the world as a contingent case. It is possible to manage systems and the environment only when the potential of cybernetic thinking is realized in language as a pragmatic form of life. Such a language becomes a visual conceptualization of the relationship between structures and functions. Since the image in the digital environment no longer represents anything, the metaphysical thinking established in philosophy, art, and sciences has been suspended. The image cannot be a function of external factors, nor does it arise from an otherworldly source, such as God's hidden presence, as in the era of myth and religion. Instead, the image depicts what occurs during the emergence of life in this phenomenon.

The connection of genes and information as the emergence of new life determines, for example, the essence of bioart and transgenic art (Paić, 2022b; Sorgner, 2022, pp. 191–219). When we have all this in mind, it should be not difficult to understand that the essence of transhumanism in the philosophical sense necessarily appears within the framework of what belongs to posthumanism, which is the realization of the nature of the technosphere in a posthuman condition. If for Nietzsche the last man was a preparation for the arrival of overman as the meaning of the Earth, then in the case of *homo kybernetes* we find ourselves in a position of thinking that can no longer be philosophical or artistic, nor scientific research into the first reason and the last purpose of the human-too-human in the age of the global-planetary development of the technosphere. Cyborgization refers to the previous instantiation of producing technologically perfected and improved "beings" that have become hybrid structures and applications of genes and information in everyday life.

Why are we talking about "the last instance" of this border crossing between humans and posthumans? Because what inevitably comes cannot be non-autonomous or non-self-regulating, it must change. What is coming as a technological singularity must necessarily be freed from any external management of the system and environment. This means that technosphere thinking requires overcoming the limits of metaphysical thinking, starting with the introduction of the concept of *autopoiesis* (Paić, 2021, pp. 431–456). As we can see, Sorgner, from the so-called realistic perspective, rejects the futuristic-utopian techniques that one strand of transhumanism explicitly promotes. These are cryonics and uploading. The body and mind are no longer considered Cartesian, but rather anti-essentialist. If the approach to the new cybernetic physicality is derived from a different relationship to the term "enhancement ", then we must distinguish between technique as a means and emergent technology as the means-end of the entire process of creating a cyborgized *posthuman condition*.

Since the invention of language at the dawn of human evolution marked the beginning of what we call animal rationale, one should be highly cautious when making assumptions about the future of language in a *posthuman condition*. The reason is that the pragmatics of *know-how* thinking only require formalized rules of use for a thing or set of events to be realized virtually. Language thereby loses its meaning or purpose, as Heidegger attributed to the telling thought of human existence (*Dichten*) (Heidegger, 2002). Instead, all humanities, and what Sorgner calls posthuman art, need a different linguistic articulation of a meaningful horizon in the interaction between system and environment. The changed character of language in the logic of new media, rather than the syntactic-semantic clarity of speech, becomes a kind of "white noise" in communication. In this way, the pictorial structure of language in the age of the technosphere, as already announced by Ludwig Wittgenstein in *Philosophical Investigations*, becomes a *language game* (*Sprachspiele*) (Wittgenstein, 1998).

What seems particularly interesting in Sorgner's consideration of the philosophical problems of transhumanism is his stance on the so-called nihilistic perspective and the concept of positive pessimism. Of course, he disclaims that all directions and tendencies in this interdisciplinary movement today are ″nihilistic, pessimistic-positive″. Still, according to the statistical analysis, they are mostly types of naturalism, atheism, or secularism (Sorgner, 2022, p. 11). Schopenhauer was the only one who took pessimism seriously, even asserting that art, in its profound ontological truth, is imbued with pain and suffering. Its sublimity becomes the last possibility of deliverance from the abyss of life's disappointment and death's agony.

But, as we know, Nietzsche assigned to art the attributes of the most magnificent possible turning point in the effort to overcome the nihilism of Western metaphysics and therefore removed any pessimism from this doctrine of self-affirmation of life. Overall, Sorgner demonstrates how apparent paradoxes and aporias arise within transhumanism at the level of "worldview". The denial of God and complete naturalism, in developing assumptions about the possibility of human transition beyond biological limits, may simultaneously lead to a precisely nihilistic attitude towards the historical-epochal dimensions of this transformation of life. Namely, although pessimism is mainly characterized by a negative attitude towards the future of the human species and its prospects, the positive one stems from trust in the power of the mind, as "common sense", crystallized in what modern technology provides. In this way, namely, it is shown that the key point of the synthesis of mind and technology is that opinion which, although Sorgner does not carry it out in his analysis and although he does not have a mostly positive attitude towards Heidegger, tries to connect the very beginning and modernity starting from the metaphysical frameworks of thought in which cybernetics was born.

Heidegger, therefore, said the following about it in the most concise way:

> Therefore, the circular rule is a fundamental principle of the cybernetic world. The possibility of self-management and automation of the movement system rests on this. In the cybernetically represented world, the distinction between automatic machines and living beings disappears. It is neutralized in the undifferentiated event of information. (...) Humans are also drawn into this monotony of the cybernetic world. And even in a distinguished way. From the point of view of cybernetic representation, humans have their place within the widest circle of rules. According to the modern idea of humans, he is, namely, a subject who relates to the world as an environment of objects by processing them. Hence, all kinds of changes in the world come back to a Human. The subject-object relationship, as cybernetically represented, is a mutual relationship of information and feedback loops within a distinct circle of rules, all of which can be described

under the title 'Human and the world'. (Heidegger in Jaeger and Lüthe 1983, p. 16)

From what Heidegger said about the essence of cybernetics, which becomes in the technosphere a planetary-global space of virtualization and implosion of information, can we arrive at the possibility that what transhumanism has as its postulate, regardless of its differences in the so-called directions, open as an *autopoiesis* in which the possibility of preserving both ″Human' and the world″ remains beyond their almost inevitable reductions? This question seems decisive for the prospects of the coming philosophy. Any distinction between humans and non-humans presupposes that a third exists (*tertium datur*). This "Big Third" throughout the history of metaphysical thinking has always been thought of as *God, Nature, causa sui, superego, sublime object X*. But isn't it astonishing that even in Hegel's science of logic, the third is nothing more than a subject-substance or a thing of thought that sets itself in motion? After all, isn't freedom precisely the autopoietic event of thought that sets in motion every corporeality, understood in one way or another, just as it sets in motion every soul-spirituality (*psyché-logos*), understood in one way or another? Isn't freedom a condition for the possibility of absolute in one way or another?

2.3 Freedom Beyond the Absolute

Freedom as absolute? This is precisely the essence of posthumanism in philosophy. Take the so-called destiny into one's own hands, transforming the necessity of historical events into the possibility of controlling contingent events in the process of becoming (*Werden, devenir*) different from the same. Almost all ardent theories of post-and transhumanism, programmatically, manifestly, and triumphantly demonstrate that, owing to the progress and development of the mind throughout history, the time has come for humans to free themselves from the shackles of the historical tragedy of their existence. Mind, therefore, is identified with knowledge, which in the form of technoscientific construction of new beings fundamentally changes the natural-biological order of relations found, which was almost unchanging for metaphysics from Plato to Hegel.

This is evidenced by statements from most technoscience representatives, who argue that, under laboratory conditions, synthetic biology has enabled the creation of an artificial being, such as the so-called oncomouse. All these points suggest the real possibility of introducing artificial life into the increasingly complex aspects of nature and human habitation on Earth, as well as beyond its natural-biological boundaries. Moreover, that event represents a turning point in understanding the distinction between the human and the posthuman, and, in the example above, also in the distinction among prehuman, human, and posthuman. History in a different, post-metaphysical sense begins with the logic of artificial intelligence and irreversibly refers to the cosmic-natural evolution of the animal-human-machine.[2]

Freedom, as the thought-life autonomy of human existence, does not contradict this event of absolute techno-genesis. This was already clear to Gilbert Simondon when he defined the essence of substance in a fundamentally different way—one that, in Cartesian philosophy, was traditionally explained in terms of matter's inertness in its expansion, the *res extensa*. We have seen that the problem with the cybernetic paradigm is that it represents, on the one hand, the realization of metaphysics, and, on the other, its transgression by introducing the principle of *tertium datur*. Matter can no longer be left to vulgar materialism in this new thinking, just as it cannot be idealized with some spiritualized model of vitalistic materialism. Matter, therefore, has not only an energy potential, without which there would be no complex network of interpenetrating bodies in motion, nor any freedom of movement of monads in a pre-established harmony à la Leibniz.

To be able to think freely in the sense of *causa sui*, so that human thinking in the form of living-and-cybernetic physicality could reach the idea of absolute self-consciousness not only of man but also of a machine as a cyborgized "creature" with a soul-spirit, it is necessary to bring together the concept of information with the idea of *autopoiesis*. In this way, two essential cybernetic reductions occur: one that refers formally to the metaphysically understood res cogitans and the other that affects the *res extensa*. This necessary anti-substantialism, which Simondon sought to realize radically in his reflections on the relationship between the Human and the machine (Simondon, 1969/1989, pp. 234–236).

Neither the mind nor the body is separated by an abyss anymore. Therefore, information is neither a form nor a substance in the Aristotelian sense, but that which interferes, which creates new information, and thus produces events as pure contingencies. When the mind in its absolute freedom thinks in the human form of self-consciousness about an *object X*, it is not mere contemplation and speculation about the Kantian thing-in-itself (*Ding-an-sich*) that affects our senses. Thinking cannot be merely the indifferent observation of a being in nature, confined to the mere objectivity of the object, on which the ideal of the Greek paradigm of knowledge about the world rested. Instead, already in the new Era, thinking becomes a pure construction of the object of cognition and, hence, a subject-substance way of producing new beings or objects in the world.

However, with the advent of the cybernetic paradigm, due to the concept of information and the possibilities of machine information processing, we are no longer dealing with something "in" the world, and we are not thinking "about" something intra-worldly or supra-worldly. *We*, as a union of natural mind and artificial intelligence, produce objects *X* that exist only when changed substances and the appearance or form of these substances are joined by technoscientific construction. This process, in which the power of the technosphere reigns auto-poetically as a new absolute, is therefore not nihilism without the meaning of history and the complete loss of the human-to-human.

On the contrary, this process, as advocated by radical transhumanism, including cryonics and uploading techniques, posits the technological singularity as a form of absolute freedom within a new mode of thinking beyond metaphysics. It appears we are encountering something that is both monstrously fascinating and almost uncanny (*Unheimlich*). Namely, we are accustomed to defining freedom neither as a Being nor as a Being, but rather as a condition for the possibility of their relationship in common existence. Thus, for example, in Schelling's account of the distinction between the essence of human freedom and human freedom itself, questions such as freedom of the will, Good and Evil, and personality arise. However, this difference is meaningful only if it is recognized that absolute causality, in essence, entails only unconditional passivity for all others.

> Through freedom, a fundamentally unlimited power is asserted alongside and beyond divine power, a conception unthinkable within these concepts. As the sun in the firmament extinguishes all the lights in the sky, even more so does the infinite extinguish every finite power. Absolute causality in One Being leaves only unconditional passivity to all others. This entails the dependence of all beings in the world on God, and that even their continued existence is only an ever-renewed creation in which the finite

> being is produced not as an undefined generality but rather as this definite, individual being with such and such thoughts, strivings, actions, and no others. (Schelling, 2006, pp. 11–12)

Suppose this position of Schelling's is understood as an essential insight into what makes freedom the abyss of human existence. It cannot be reduced to Aristotle's definition of man as an *animal rationale*. In that case, we are faced with something even more monstrous than what Schelling himself could have anticipated when he discussed free will, Good and Evil, personality, and so on. Why? If, for Simondon, the robot does not exist but has a different mode of Being, then, from the perspective of absolute metaphysics, we should examine how human freedom is transmitted to the possibility of non-human freedom in the posthuman condition. All beings in the world constitute a living unity, a synthesis of physical, mental, and spiritual aspects, as well as natural, biological, and autopoietic elements, understood in terms of homeostasis.

2.4 Conclusion

When, therefore, Simondon assumes in his defence of the cybernetically understood coexistence of animal, Human, and machine a kind of secularized idea of the Good beyond all pantheism and atheism, he is talking about the belief that progress and development in history are only possible from the pre-established harmony of the organic and the mechanical. But what if, for the understanding of the essence of the technosphere as absolute freedom of self-movement and self-management of life, it might be impossible to neutralize and suspend what is the negation of freedom of will, and rests on, namely, evil in its pure contingency? What, then, if freedom in its absolute construction of events is indifferent to good and evil, and shows itself only in the pragmatic form of one's self-consciousness that combines the living and the artificial in something that formally has the characteristics of evil and destructive tendencies against an evolutionarily lower stage of development in the form of animals and humans?

The renowned astrophysicist Stephen Hawking warned about this near the end of his life, expressing his concern about uncontrolled progress and the development of artificial intelligence. We now see that the philosophical problem posed by post- and transhumanism is precisely the one posed by Schelling through the speculative thinking of German idealism. This is the problem of freedom as an absolute that starts from unconditional causality and thus subjects everything else to unconditional passivity. The technosphere, as an autopoietic system of absolute freedom in information production, with a tendency-latency, also represents a system of information control and, as such, has the same logic of action as the concept of "human freedom" in mental action. However, this is not the essence of freedom, because information cannot be reduced to the logic of organically understood causality and, accordingly, to teleology, i.e., the ultimate purpose of all events. Freedom cannot be either a Being or a being, and its ″essence″ is not reduced to absolute causality, but the event of an onto-cybernetic network of pure contingencies. Evil is a first-class philosophical problem because it connects Being, God, the world, and man. However, evil is neither "in" nor "outside" of the world, but within it as a constellation of relationships that, in some contexts and situations, creates evil creatures and violence as a threat to the destruction of the world in general.

How then should it be possible to overcome that which affects the fate of the world, which Human cannot abolish, and which was considered a tragic event by the Greeks since ancient times, precisely because a man had the freedom of will to autonomously choose his destiny, like, for example, King Oedipus, who defiled his mother and killed his father? It will be said that he did it unconsciously, unaware of the causes and consequences that led to the tragic act and the subsequent realization of the truth. That's precisely the problem. Unconsciousness and ignorance are excluded from the logic of autopoietic thinking systems. When information interferes with the interactors in communication, it may create a metastable equilibrium, as Simondon calls it. What is the meaning of that attitude except that, paradoxically, the unstable in the form of new chaos now enables the system to balance, and this should be only possible when entropy, as the second law of thermodynamics, regulates processes in the cybernetically-based relationship between Human and the world.

The result is that we do not rule the environment in the sense of absolute control over evil as an event capable of destroying not only the biosphere but also the technosphere as such. If it seems to someone that this could be a post-metaphysical ″Big narrative″ about the *war of the worlds*, then they might be truly right. The philosophical problem with the emergence of the posthuman condition is not in any relationship between humans and non-humans, but in how freedom, in its essence, opens in the realized world of the technosphere, where the logic of absolute causality and teleology no longer rules. Instead, we encounter a metastable game of chance in which pure contingencies produce the coming that is never devoid of the possibility of being unpredictable and beyond the control of the human mind. *Homo kybernetes* denotes the final stage of human evolution and, at the same time, the culmination of metaphysics in cybernetics. It only means that, for the first time in history, freedom has become a contingent necessity of thought, not always an eternal fate assigned by the gods or God as a reward and punishment for man's dignity as an irreducible being under the stars.

2.5 Notes

1. If for Simondon, the machine in the cybernetic "third order" appears precisely as a self-propelled and self-managed mechanism of purposeful action, and Human has only the function of overseer and mediator of the process of the autonomous "concert of machines", it should be evident that there can no longer be the talk of superiority-subordination in terms of the articulation of power in the physical-biological-psychosocial framework. Therefore, humans and machines are no longer determined from a metaphysical perspective, but from a cybernetic set of relationships between living systems and the environment. Living systems are possible in the technical world, but only when they incorporate technical objects in the same way as other living beings, such as plants and animals. When onto-genesis becomes techno-genesis, we enter a period of the disappearance of all previous standards of becoming a "Being", "beings", and "essence" of Human. Anthropology that adequately corresponds to Simondon's utopian demand set back in the 1950s by the "philosophy of individuation" no longer corresponds to what this term meant for Kant, and even less so for many social scientists (sociologists and ethnologists) during the twentieth century. Instead of the science of Human, which, in one way or another, logocentric or anti-essentialist, still considers man "the measure of all things", Simondon proposes concretization (*concretude*) or the openness of the machine itself as a condition for the possibility of a new homeostatic, metastable, informational balance (Simondon, 1994).

2. "The most important reorientations of recent science do not arise at the methodological level but at the ontological one. It concerns the "techno-scientific turn" in the context of scientific research objects. Many objects of scientific study are no longer part of the untamed natural world; they are produced artificially. The technoscientific turn at the ontological level means that most of the entities and processes addressed by scientific research are not discovered but created. Laser processes, chlorofluorocarbons (CFCs), nanotubes, or layered systems of ferromagnetic materials and chromium in a magnetic field (in which giant magnetoresistance makes its appearance) would have hardly ever come to exist by nature left to its own devices. They are deliberately produced by humans but explored, like objects generated by the whims of nature" (Carrier in Nordmann, Radder and Schiemann (eds.), 2011, pp. 51–52).

References

Carrier, M. (2011). "Knowledge is a power," or how to capture the relationship between science and technoscience. In A. Nordmann, H. Radder, & G. Schiemann (Eds.), *Science transformed? Debating claims of an epochal break* (pp. 43–53). Pittsburgh University Press.

Deleuze, G., & Guattari, F. (2005). *Qu'est-ce que la philosophie?* Minuit. (Original work published 1991)

Haraway, D. (1990). *Simians, cyborgs, and women: The reinvention of nature.* Routledge.

Heidegger, M. (1983). Die Herkunft der Kunst und die Bestimmung des Denkens (1967). In P. Jaeger & R. Lüthe (Eds.), *Distanz und Nähe: Reflexionen und Analysen zur Kunst und Gegenwart* (pp. 11–22). Königshausen + Neumann.

Heidegger, M. (2002). *Was heisst Denken?* V. Klostermann.

Heidegger, M. (2018). *Sein und Zeit* (2. Aufl., Vol.GA. 2). V. Klostermann.

Nietzsche, F. W. (1999). *Also sprach Zarathustra. Ein Buch für alle und keinen. Sämtliche Werke.* KSA. W. de Gruyter.

Paić, Ž. (2018–2019). *Technosphere* (Vols. I–V). Sandorf and Mizantrop.

Paić, Ž. (2021). Thinking and creation: Arché-Ergon-autopoiesis. In *Nihilism and contemporaneity—In Nietzsche's footsteps* (pp. 431–456). Litteris.

Paić, Ž. (2022a). *Art and the technosphere: The platforms of strings.* Cambridge Scholars Publishing.

Paić, Ž. (2022b). Metaphysics and cybernetics: About the technosphere or from the thing of thought to the thing that thinks. In Ž. Paić (Ed.), *The technosphere as a new aesthetic* (pp. 1–36). Cambridge Scholars Publishing.
Rieger, S. (2003). *Kybernetische Anthropologie: Eine Geschichte der Virtualität*. Suhrkamp.
Schelling, F. W. J. (2006). *Philosophical investigations into the essence of human freedom* (J. Love & J. Schmidt, Trans.). SUNY.
Severino, E. (2016). *The essence of nihilism*. Verso.
Simondon, G. (1969/1989). *Du mode d'existence des objets techniques*. Aubier. (Original work published 1958)
Simondon, G. (1994). *Une pensée de l'individuation et de la technique*. Chatelet.
Simondon, G. (2005). *L'Individuation à la lumière des notions de forme et d'information*. Jérôme Millon.
Sorgner, S. L. (2020). *On transhumanism*. Penn State University Press.
Sorgner, S. L. (2022). *Philosophy of posthuman art*. Schwabe Verlag.
Stiegler, B. (2018). *The neganthropocene*. Open Humanities Press.
Wittgenstein, L. (1998). *Philosophical investigations* (4th ed.). Wiley-Blackwell.

CHAPTER 3

The Technosphere and Nihilism: Autopoiesis as the End of Metaphysics

3.1 Arché

The original or initial (*arché*) is considered far-reaching and decisive for the understanding of Being. This applies equally to the mythical and religious determination of the origin and becoming of the world. The Greeks begin with the story of the gods' origin and their role in theogony, and the Bible reveals that the absolute creator of the world, Yahweh, shapes the universe in seven days, which, cosmologically, has the power to establish the earth and its creatures. This Human has a position and place of mediating with the divine light of eternal presence. It is not just a connection between creation and time, serving as a dividing line between what belongs to eternity. There is still something before the emergence of Being, beings, and being human in the historical context. And that "something" becomes, of course, nothing but the moment of the emergence of thinking as an event.

With it, all that "is" takes on the possibility of a synthesis of Being and time beyond nothingness and emptiness. The original or initial (*arché*) for metaphysics is therefore straight to the event, without which nothing can be thought. Moreover, it seems that the very first, as well as the One and the singular "that" which comes from the abyss of Being and time, is not only the mystery of the origin of Being in general and as such, but above all the mystery of thought as "other creation". What is secondary or different always denotes the product of difference and stems from the

Ž. Paić, *The Technoscientific Turn of Philosophy*,
https://doi.org/10.1007/978-3-032-21081-4_3

identity of Being. To create in the way of artistic production of worlds (*poiesis*) means, therefore, necessarily to imitate the reproduced "original" in the sense of Being as nature (*physis* in Greek, *natura naturans* and *natura naturata* in the modern understanding, for example, Spinoza) .(Heidegger, 1998, pp. 10–11)

The second creation cannot be credible if it does not bring the "new" into the world at the same time. This is true even when it seemingly mimics the divine creation of nature in a "realistic" manner. The distinction between the original or the initial (*arché*) and that which is secondary and different proves essential for the ontological determination not only of art, architecture, and design but also for the essence of philosophy and science. Suppose the difference is always what presupposes the same, or the identity of Being with itself. In that case, its provision is that it must always be in a state of difference and otherness with respect to the same. But suppose its essence cannot be derived from Being but is a difference as such, or one that Deleuze claims derives from itself. In that case, we encounter a grounding of metaphysics and its fundamental structures and conceptual-categorical relations (Deleuze, 2011).

The question of the first or initial (*arché*) cannot be answered by the positive sciences, such as mathematics and physics, nor by modern cosmology and astrophysics. The reason lies in the fact that the original or the initial (*arché*) cannot be hidden in the essence of the number with which the infinite sequence begins. But it also cannot be resolved by cosmological speculation about the distribution of matter, energy, and information. Why? For the simple reason that it is the original and initial (*arché*) ontological problem *par excellence*, which for metaphysical understanding denotes at the same time a question of the squareness, which for Heidegger represents the structural unity of Western metaphysics in general: Being-God-World-Human. To think means to always think within this historically formed framework or paradigm, even when, as in the case of radical deconstructors of metaphysics like Nietzsche and Deleuze, they resort to the abolition of God as the authoritative concept and idea. Thinking does not mean merely presenting something as something, nor can it be reduced to the volitional-perceptual processes used today by neurocognitive scientists to convince us that thought always represents a will and action beyond the traditional relationship between subject and object.

To think means to be focused on the purity of the creation of thought as the production of ideas and conceptual formations on the event, with which Being and time become the task of thought. Heidegger said all this during his own "journey of thought" as a "task of thinking". If, therefore, what the matter of reasoning requires proves to be a relation between the original or the initial (*arché*) with the secondary or different from the same. An essential problem of all metaphysics already opens up, consisting in the fact that every creation presupposes the idea of creation: or as what is the first cause and last purpose of Being in general, or as what transcends such a circular structure of thought and reaches for the way out of that "fateful" fourth with which Heidegger ended his thought journey. But instead of an onto-theo-cosmo-anthropological structure, the path ends with the fourth of earth and sky, mortal and immortal, as a path to the "other beginning" (*andere Anfang*) (Heidegger, 2005).

Getting out of the "vicious circle" does not mean abandoning metaphysics as the execution of some arbitrary decision. The necessity by which metaphysics is established throughout the history of the West, both at its beginning and at its end, does not mean that thought becomes merely the realization of its "laws" and "forms" of utterance in language as the abode of *logos* and *mythos*. Necessity denotes only that which is, admittedly, different from possibility as difference. But it proves to be the last possibility of some thought act in the realization of the purpose of man's existence. What is necessary could never be so inevitable, for the primary or initial (*arché*) is the possibility of thinking in the openness of the horizon within which the telling of a Being becomes a measure of one's own time. We cannot, therefore, leave the soil of the earth from which we think "about" the world and "on" Being as such, because the language by which we always and always think is "necessarily" determined by a rootedness in the homeland and its periodic shadows of life.

What arises and opens as original or initial (*arché*) presents an ontological puzzle. How do we come to think of something as original or initial (*arché*) if we also start from the idea that "what is" in the original sense is uncreated and eternal, as Parmenides puts it? Is it not logically contradictory, namely, to speak of the original or the initial (*arché*), if Being as such and as a whole has neither a beginning nor an end because in its permanence it is unchangeable and eternal? Dealing with the issue of the end of metaphysics from the original question of the meaning (or truth) of Being, Heidegger, in his lectures on the beginning of Western philosophy in Greece, began with the reflection of the "first philosophers", primarily Anaximander and Parmenides, on the "first beginning". In doing so, he credibly linked the origin or beginning of philosophy with what he called the "end of philosophy" in the realization of the possibilities of metaphysics as such.

This connection denotes, at the same time, a question of the possibility-reality-necessity of a philosophy that begins with the thought of Being and ends with the idea of cybernetics as the realization of metaphysics in a technically determined "world". In short, Heidegger's consideration of the issue of beginning cannot be merely a question of the "necessity" of philosophy as the beginning of Western history, but above all a question of the true possibilities of the historical event of Being itself in its original time. In this sense, philosophy cannot be separated from its own "mission" and "necessity" as the essence of Western history, marked by the "first beginning" (*arché*) and "last purpose" (*telos*). However, history can still reveal something about its condition. This should undoubtedly be the first and last aspiration to merge with Being and beings as a whole. The archaic age can never be relegated to an irreversible past. It seems alive only when there is an urgent spiritual need for "origin".

3.2 Appearance

In the above lectures from the summer semester in Freiburg in 1932, in β 6. Heidegger expounds on the ruling origin of Being as the enabling power of appearance. In doing so, he refers to Anaximander's saying *arché ton onton to apeiron*. In free translation, Heidegger says that the meaning of this fragment is that it is the origin or source of Being as such, concerning infinite being (*das Grenzenlose*) (Heidegger, 2012, p. 27). The following explanation shows why the term being should be used as such, and not as an individual being. It is not, therefore, a "coincidence", a mere singularity of Being in a multitude of appearances, but an internal connection/relationship of Being and being. In doing so, it should be evident that Being is not superior to beings in the theological sense, where God rules and governs the world, and beings serve Him submissively as stewards of His absolute will.

The Greeks, with the notion of the origin, source, or beginning of Being as such (*arché*), understood it as arising from the free relationship between Being and beings, rather than from a framework of thought that presupposes a complete hierarchy of being. Hence, this *arché*, the archaic in itself, opens and enables the appearance of Being as that which is boundless as such. When we consider that Anaximander's saying or fragment offers a comprehensive insight into the primordial understanding of the relationship between Being and beings, it becomes evident that the relationship between *apeiron* and *arché* should be mediated by the need for connection between Being and beings. The Greeks denote this concept by the term *chreon*, which Heidegger translates as Not (meaning necessity or destiny) in German (Piementel, 2019, p. 13). What is "necessary" and "fatefully" determined here? Nothing but th*e* relationship/connection between what is original or initial (*arché*) and the way the being appears as such.

> *... Arché is what precedes everything and from which everything derives. This is about the beginning of Being, the appearance of...* (Heidegger, 2012, p. 27)

Appearance, however, must not be equated with something that belongs only to the world of phenomena in contrast to what is beyond any appearance. Thus, the *arché* in Greek thought cannot be the *a priori* that grounds the *a posteriori*, as is customary in Kant's transcendental metaphysics. Heidegger explicitly argues that this is not the beginning of the appearance of beings, but a Being as a whole. Hence, Anaximander's saying about the beginning of what appears as the preceding and originating at the same time denotes a pre-metaphysical "credo" of Greek philosophy.

To think of the primordial beginning of the appearance of Being means to reveal the horizon of the appearance itself as such. It is therefore unsurprising that Heidegger regards these instructive and "fateful" words and concepts, such as arché, as the beginning of all Western history. Of course, history is inconceivable without thought, just as thinking becomes inconceivable without the historical occurrence of Being in time. This mutual familiarity of Being and thinking reveals how the narration in Greece unfolds as an attempt to "establish" the world in its original appearance, free from all other determinations. All words or sayings about the Being derive, therefore, from this source/origin of the Being as such and as a whole. The idea of a common source of Being seems undoubtedly what Reiner Schürmann calls the "broken hegemony" of Western metaphysics (Schürmann, 2003).

The primacy in the ontological sense of the idea of the beginning (*arché*) of a Being does not mean that its essence lies outside Being and that Being is superior to that which is subordinate in the sense of absolute command. Archaic cannot be anything "archaic", in the mind of the irreversible past, because primordial or original time is timed precisely in the unity and harmony of the three dimensions—past, present, and future. What was considered "original" at the beginning of Being never really passed. The primacy of the "original" signifies what was as evident to the Greeks as to all archaic peoples worldwide: namely, that the past, as Being, denotes the temporal dimension that physically determines the meaning of Being in general. In the sign of the rule of what this word signifies, Heidegger may therefore say that.

> *Arché*, as the ruling origin, remains present in everything, coming first and ultimately manifesting in all its forms and disappearances. (Heidegger, 2012, p. 28)

Let us add to the saying that Heidegger uses the German word *Verfügung* instead of the Latin word *principium* to give the Greek term *arché* an appropriate meaning. And this, as we have seen, is both principle and order and direction and command in the sense of the rule of the original and "founding" even in the disappearance of appearance. This seems important to note because the meanings of the basic Greek words for philosophy and metaphysics extend beyond metaphysics. Language and speaking cannot be reduced to a mere exchange of information. Speech that combines the apophantic and performative meaning of narration at the same time refers to an event by which Being and time are understood pre-ontologically, on the border of silence and ineffability. To that extent, the discourse on the initial and the original, the principled and the original, that which commands the course of history in a commanding manner, and when this does not seem so visible, needs to enable the emergence of Being as a whole. Orientation cannot be merely a display of the path from beginning to end, with its causes and purposes.

In what enables direction lies the origin of the path of thought, and this is what Heidegger derives from the primordial narration of the Greeks, as found in Anaximander and Parmenides. *Arché* opens the possibility of the appearance of beings in the light of the meaning (truth) of Being and is therefore inextricably present throughout the history of Western thought, or philosophy, as metaphysics, culminating in cybernetics. To be contemporary cannot be enough; it is not enough to live in modern times, guided by the rule of enframing (*Gestell*). It seems necessary to understand time as a connection or relationship between the initial and directing course that has never really passed. It is that time that does not last but is timed in memory as the existential relation of the essence of Human to Being in general. The past, therefore, acquires its essential meaning only from the future as an impending event. But the true future connects the past and the present into an open event of the world to come, within which Being gains meaning in general.

3.3 Conatus

In his lectures on the beginnings of Western philosophy, Heidegger exposed the unprecedented "modernity" of the trinity of reference words in Anaximander: *chronos, arché*, and *apeiron*. In the twentieth century, the Greek language became ambiguously mythopoetic; "Alive" is its merit, and after Heidegger, this hermeneutic path must be continued by Hans-Georg Gadamer. Of course, traces of this path in the nineteenth century were outlined primarily in Nietzsche's thought and Hölderlin's poetic mythology. However, thanks only to Heidegger, what shone with full splendour is the attempt to explain the accurate contemporaneity of our time, marked by scientific and technological innovations, by turning towards the "beginning" (*arché*). In this, we must see the complete "novelty" of philosophizing beyond Hegel's dialectic and Husserlian phenomenology, not to mention the various neo-Kantian renewals of the notion of the subject, or the metaphysical positivism in multiple directions.

When, therefore, we establish a living relationship with the Greek experience of thinking of Being before Plato and Aristotle, then the main reason for this "long journey through the night" denotes the never-gone past in an attempt to regain what was lost over time, which means the erosion of the metaphysical soil of history—remained deeply repressed into the abyss of memory. Neither Anaximander nor Parmenides is in themselves "modern" without the mediation of Heidegger, Gadamer, Severino, Schürmann, and other thinkers of our time. On the contrary, their place cannot be determined only by "newer" philological-philosophical contributions. More significant should be something else: how can this "deep" reading of sources and origins be understood as the experience of thought after the "end of metaphysics"? Does it have any other prospects than those set out in the postulate that the ancient past has an entirely different meaning for contemporaneity than the "dead tradition" of conservatism of all colours?

Let us dwell on the discussion of the orientation and rule of the beginning (*arché*) on Heidegger's postulates about Anaximander. It should be said at once that the fundamental reason for this reckoning with the "forgetting of tradition" lies in the notion of the meaning of Being, without which it becomes impossible to emancipate authentic history from history as a scientific-positivist approach to the past. With this in mind, it should be obvious why contemporary thinking, in the wake of Heidegger, "seeks" a different understanding of the essence of philosophy in general than that which belongs to the metaphysical tradition from Plato to Hegel. Heidegger persistently "translates" Greek classical texts in much the same way as Nietzsche, contrary to the mainstream of modern philology, which is dedicated to Greek pre-Socratic thought.

We will not engage in these controversies here, as they are not a topic for us to consider. However, it should be noted that Heidegger's way of approaching the Greeks is nothing but "credibly unreliable", i.e., aimed at trying to dissolve through the German language from Meister Eckhart to Hölderlin and Nietzsche which in the narrative bears a trace of non-metaphysical origin of history, which is precisely because it is alive because it cannot be reduced to the fossils of the positive nihilism of the humanities of modernity. Therefore, this only means that telling about Anaximander, for example, must pave the way for thought that speaks an ancient language, often on the border of myth and poetic ambiguity, to open the possibility for philosophy and art, that is, for thinking, creating, and aspiring to meet the event (*Ereignis*) of the "second beginning". If, in the "first beginning", the event was given without human participation, it is now up to him to "give birth" to new gods of "self-creative chaos" from the essence of philosophy and art.

Nothing happens without this unique *conatus*. Thought seeks to overcome obstacles along the way, free itself from the siren call of realism and idealism, and find in the singularity of the conceptual speech of the thinker a safe place of mediation without external intermediaries. Thus, thinking as a philosophical path focuses on a fundamentally different mode of creation (*poiesis* vs. *creatio*) from that of art. However, to a large extent, since Nietzsche, their discourses have overlapped.

Why, then, so much almost "messianic" expectation from the opinions of early Greek thinkers with their instructive philosophical words—directions that last for centuries and define our thought life as the only real life? For Heidegger, Anaximander's statement about the beginnings or emergence of Being as such (*arché*) is at the same time a statement that everything originates from the source and returns there by the nature of the thing (thinking) itself. What does this mean for contemporaneity, and does it have any deeper meaning in the age of the rule of astrophysics and theories of singularity and contingency? First of all, Heidegger reads from it the necessity of the "eschatology of the West" because Being is thought starting from this beginning, which is indestructible and uncreated, eternal and permanent (Heidegger, 2003a, 2003b, p. 327).

However, the eschatological denotes nothing but "necessary" arising from Being and hence "fateful" in a completely different sense from the fulfilment of a predestined mission in the sign of God's election of grace, as it appears in the theological heritage of Christianity. To think of Being as an eschatological mission means to consider the direction of what was once created and never disappeared, except that it was pushed into oblivion by history, thanks to the rule of historicism as one version of modern nihilism.

3.4 Metaphysical Circle

What does an ancient maxim at the beginning of Western philosophy mean for contemporary thought? Surely, its meaning becomes apparent only when it becomes clear that the time of today's technological nihilism originates from the notion of novelty in the new from the perspective of actuality. Returning to the past from there can never be mere nostalgia. It is a resistance to the time of complete reduction of the world to the effects of "benefits" and "pragmatics". Heidegger seeks to expose the riddle of Anaximander's saying precisely because recalling the beginnings of Western philosophy necessitates a reorientation of understanding the essence of contemporaneity. The philosophical perspective that elucidates this riddle is undeniable, although its proximity to artistic creation is equally evident. At one of the decisive points in the discussion of "The Anaximander fragment" from 1946, we come across the following assumptions:

> Only in thoughtful dialogue with what it says can this fragment of thinking be translated. However, thinking is poetizing, and indeed more than one kind of poetizing, more than poetry and song. Thinking of Being is the original way of poetizing. Language first comes to language, i.e., into its

> essence, in thinking. Thinking says what the truth of Being dictates; it is the original diet. Thinking is primordial poetry, before all poesy, but also before the poetics of art, since art shapes its work within the realm of language. All poetizing, in this broader sense, and also in the narrower sense of the poetic, is in its ground a thinking. The poetizing essence of thinking preserves the sway of Being's truth. Because it poetizes as it thinks, the translation which wishes to let the oldest fragment of thinking itself speak necessarily appears violent. (Heidegger, 1975, p. 19)

We now see how important it is, both within and outside the metaphysical circle, to preserve the memory of the beginning (*arché*). Things never change in a way that makes them entirely opposed to their emerging essence. Moreover, the origin or source of the Being shows us this simple "necessity" as the first and last "possibility" of all change. Thinking in the moments of the "first beginning" is not divided into systems and classifications. It belongs to the narrative-poetic essence of language because it is expressed only through speech that conveys meaning in dialogue with the governors of the utterance itself.

Poetry and the poetic, as well as the aesthetic and the synaesthetic, come only after the event of thought has spoken the language-speech of the "original way of poetry". It is therefore not at all surprising that the Greeks, in the origin of Western thought, could not strictly distinguish between philosophy and art, that is, that which belongs to wisdom (*to sophon*) and that which appears as a creative act of idleness in poetry (speech and singing). In both cases, conditionally speaking, thinking seems to be a poetic rendering of the meaning (or truth) of Being. The immediacy of the approach to the openness of the meaning (truth) of Being, as Heidegger puts it, has the power to generate the world. In the case of Anaximander, three fundamental words—*chreon, arché,* and *apeiron*—refer to the permanence, necessity, and infinity of Being and beings as a whole. Therefore, what we call today, thanks primarily to the Christian tradition and the theological opinion of the revelation of God, the "idea of creation" is impossible to divide between the sacred and the secular, because in the pre-Socratic era, from myth to philosophy, art itself sings and paints in poetry, painting, and sculpture.

What is puzzling here concerns the relationship between the beginning and the origin (arché), and between the origin of Being and beings as a whole (*genesis*). In his discussion of the Saying of Anaximander, Heidegger explains in detail how the concepts of this early Greek thought developed from the inner necessity of the ontological complex with which philosophy begins. However, philosophy cannot be superior to language, as in narrative poetry (*Dichten*). What must amaze us as contemporaries of the technosphere (artificial intelligence and gene manipulation) is that in an almost "primitive" form of narration, Anaximander not only made cosmologically important speculation about the nature of Being and beings as a whole, but also "founded" a way of thinking that unites and separates philosophy and art.

From the word as an idea, which *arché* is in a far-reaching sense, not only the beginning of Being and the whole of Being as such develop. There is something else that grounds the relationship between philosophy and art, creating a relationship of necessity and an unbreakable bond. It is, of course, the mother of all the arts in the classical understanding of the construction and construction of the world—*architecture*. However, it is already associated with the definition that Aristotle attributes to human beings in the *Nicomachean Ethics*. It is a work, activity (*ergon*) in the form and materiality of a job that combines technical art and poetic or aesthetic action in shaping the world (Agamben, 2019, p. 7). Archaic peoples are adorned with a commitment to building in a cult festival precisely those "works" that have the character of permanence and aspiration towards eternity. Therefore, the Greeks lacked a coherent aesthetics or theory of art as a reflective activity that gives works their meaning and significance in the world. Architecture is like the narrative poetics of thought, the initial or the final. With it, a form of human existence emerges as the realization of the possibility of the work itself in practice.

Construction does not serve the external purposes of human habitation, even when fully functional, as in modern skyscrapers and railway terminals. And they, we will admit, really have nothing of the traces of the mythical-religious past when architecture still glorified the gods and God in the construction of temples and cathedrals. Its essence lies in the relationship between *arché* and *ergon*. In fulfilling the possibilities of the human being as its existential necessity of dwelling in the world, the creative act itself becomes an event. The difference between artistic urban planning and functional city planning, according to the model of the machine of mechanical reproduction (Camilo Sitte vs. Le Corbusier), becomes only an ambiguity in the historical sequence by which architecture "lands" and the governor of the sacred becomes a means to a purpose. What should be decisive for the event of this "necessary" alienation and realization of life itself, which becomes part of absolute construction, is that at the end of metaphysics, in the age of realized nihilism of architecture and art, creation itself becomes an opinion about the act of creation. Sloterdijk, in his treatise on Nietzsche, in which he finally speaks of the cybernetic corporeality of the world, calls this the mythopoetic "eye with which Dionysus observes himself" (Sloterdijk, 1986, p. 170).

3.5 Autopoiesis

Can thinking open the way to create without the beginning (*arché*) and work as a work in the form of a created thing (*ergon*), reaching what determines the essence of the technosphere as a realized set of calculations, planning, and construction in the idea of self-organization, self-government, self-movement, things that become a thinking object (*autopoiesis*)? We have seen that, in the origins of Greek pre-Socratic thought, as exemplified by Anaximander, the poetic saying of language gave rise to three mutually coordinated ideas for the origins of philosophy and art: *chreon, arché, and apeiron*. With Aristotle, by contrast, metaphysics is organized around the principles of the activity of Being (*ergon*) in terms of *dynamis* and *energeia*. The third, the monstrously inhuman and at the same time the only "necessary" for creation from the original act of the divine and the natural as infinite (*apeiron*) to become possible as human self-activity in artistic production requires a connection/relationship of poetry as narration and art as the creation of the new (*poiesis* and *téchne*).

The event that leads to the creation of the "new" by uniting imagination and technique is no longer anything purely human and cannot be deduced from the logic of Being as primordial nature (*physis*). The pre-Socratic opinions of Anaximander and Parmenides no longer seem to be a guide for us here. Recall that in Homer's *Iliad*, that monstrous wooden horse, built as the "cunning of the mind" of the wise Odysseus to deceive the Trojans and allow the Achaeans a final victory in an exhaustingly long war, is nothing more than a fascinating technical construction with an aesthetic addition of form (*eidos*) horse (*mimesis*). Understanding creation from the essence of art presupposes, therefore, an insight into the nature of this connection/relationship of the poetic and the technical, imagination, and invention. Creation as the technical production of the "new" means the introduction of the "third", which transcends the human mind and organically understood nature.

In Greek mythology, these are creatures that combine disparate, opposing substances, such as the Cyclopes and the Centaurs. For the technical construction of the "new" to appear in the world at all, it would be necessary to reach the inhuman and the uncanny, which only allows creation to take on the characteristics of the human-inhuman act of shaping the "Big Third" in contrast to the divine and natural. It is a notion of hybridity or a process of hybridization. What arises and becomes thanks to the act of technical construction is no longer just what Maurice Blanchot calls in Kantian terms "sculpture glorifies marble" (Blanchot, 2015, p. 274). Still, it is a matter of transcending the organic and the inorganic, the living and the inanimate.

What was clear to archaic peoples, and especially to the Greeks, is often distorted in the understanding of artistic creation: that it is at once inhuman and frightening, like an encounter with the invisible and inexpressible (*Unheimlich*), and that this is why it transforms into a demand for human-to-human interaction. The experience of this contact with the incomprehensible and the boundless arises from the gap between the divine and the human. It can in no way be resolved by the Enlightenment model of escaping into the harmony of nature and the infallibility of rational relation to the world modelled on the natural sciences, especially mathematics and physics. No one has described it better than Nietzsche. In the booklet of prefaces for unwritten books from the *Legacy* (1870–1873), in one such preface entitled "Homer's Competition", we read at the outset:

> If we talk about inhumanity, then at the core of it lies the idea that this could be what separates man from nature. But there is no such separation: "natural properties" and those called 'human' are inseparably fused. Man, in his highest and noblest powers, is entirely natural and bears on him his eerie double character. His fearful and inhuman abilities may even be the fertile ground from which all humanity can grow in inclinations, deeds, and works. (Nietzsche, 1996, p. 61)

Creation and creator are the titanic powers that confront the relationship between the inhuman and the human. Nietzsche is right when he takes the Greeks as witnesses to the emergence of a unique and unrepeatable world that remains a model for all future generations of "humanity" in childish wonder, harmony, and an unheard-of aspiration to master history through creative madness (Dionysian-Apollonian conflict). All this opens the space to the metaphysical splendour of life itself. But still, he was wrong about one thing. The Greeks, however, coined a term for the technically monstrous, for the construction of the world from "Dionysus eye with which he observes himself". But they could not transcend the boundaries of that infinite in which their world was the abode of gods, men, and hybrid beings created by imagination. What was lacking in this pastoral symphony of harmony and nobility of soul was born with the modern turn of metaphysics into the rule of thought as the subjective calculability of Being in the mode of representation (*repraesentatio*). Heidegger, in the 1938 treatise The Age of the World Picture (Die Zeit des Weltbildes), argued that a world becomes a picture in the sense of a subject on the stage, representing Being as such (Heidegger, 2003a, 2003b, p. 91 (75–113)).

We have already said that Nietzsche's interpretation of pre-Socratic thinking in Greece was undoubtedly a signpost for Heidegger as well. In this book, we have demonstrated the existence of similarities and subtle differences in understanding the relationship between nihilism, as the essence of metaphysics, and the effort to overcome this inescapable necessity, as well as the "severe disease" that has plagued Western history. It is not uncommon for Heidegger to bring together, in the lectures and in the discussion of Anaximander's saying, the *arché* and the setting of the will to power as the eternal recurrence. But what seems particularly important in this case is that Heidegger saw more decisively that the problem was not in the artistic management of human existence beyond any reduction to the positivity of metaphysics in science and modern philosophy, which tends to be closed to Kant and Hegel.

A form of a system of an absolute science of spirit (transcendentalism of history) becomes a key concept at the end of metaphysics. Thus, his diagnosis of the assemblage of things in relation to nihilism is the most pervasive in modern times. The reason lies in the fact that the so-called representational thought, which begins with Descartes and continues with Leibniz and Spinoza, culminates in contemporary metaphysics of subjectivity, establishing pictorial creations as an inner possibility of different thought. This is precisely the opinion that Heidegger expresses in his book *Was heißt Denken?* from 1951 to 1952, referring to "counting" (*Rechnen*) as opposed to "poetic narration" (*Dichten*). The first includes scientific thinking, modelled on mathematics and physics, while the second encompasses philosophy and art (Heidegger, 2002).

Images result from calculation, planning, and construction. This triad determines the essence of the technosphere. However, an opinion that is no longer determined by the "necessity" and "destiny" of the first beginning and is no longer limited by what Aristotle called deed and activity (*ergon*), radically changes the whole of modern nihilism, primarily in such a way that now the very thing of thinking becomes an object that thinks. In cybernetic system theory, this concept is referred to as *autopoiesis* (Paić, 2018, p. 155–166).

3.6 Conclusion

What makes the fundamental definition of this turn in the essence of metaphysics? Nothing but the reversal of the whole structure on which metaphysics rested: in the idea of foundation/reason, everything that both Nietzsche and Heidegger explained in thinking about pre-Socratic philosophy, but mainly in the analysis of the metaphysical structure of Being, in general, starting from Plato's keywords and concepts. When the *matter of thought* is shown to be an object of thought itself, then the world, instead of being the result of onto-theology and cosmology, becomes the outcome of cybernetic anthropology. The production of virtual worlds can no longer be considered an "illusion", nor can a "simulacrum" be regarded as an illusion of some powerful reality that governs the world through God's or human transcendence. By contrast, autopoietic systems govern the world independently and internally through information, feedback, control, and communication.

Things are no longer "in" space. Instead, spaciousness constitutes a dynamic category of the absolute techno-genetic construction of the inhuman. With the help of "artificial intelligence"or the new thinking of cybernetic machines, life becomes a singular multitude. It is the event best described by Deleuze's "ontology of multiplicity and difference". On the whole, what is self-produced and self-governing brings philosophy and art to the wall of time. In other words, the fundamental question for the future cannot be longer about the nihilism of modern technology and its completion in the idea of cybernetics, as summarized by the late Heidegger, especially in his lecture "The End of Philosophy and the Task of Thought" ("Das Ende der Philosophie und der Aufgabe des Denkens") from 1964 (Heidegger, 2007).

Instead, everything should be redirected to the future, focusing on the development of "computational planning". With the disappearance of uncertainty and the neutralization of the secret in what is to come, an opinion that only creates its object is no longer a representative opinion. It has become self-presenting and self-conscious within what is called the "third order of cybernetics". Machines that think what they produce themselves are nothing but objects of thought, as images of the technosphere. From *cosmos* and *mundus*, the world has reached endless techno-genesis, characterized by the pure and indifferent self-production of ready-to-use *information*.

The circle is closed. The first and the initial revolve in the last and the final as the "eternal recurrence". But it is no longer Nietzsche's most difficult thought with which the overcoming of metaphysics truly begins, as a praise of the creative act of art that does not need any divine and human "foundation". Now it is a "matter of thinking" in the creative destiny of the world itself, left at the mercy of the technosphere. There is no going back to the beginning. All that remains is the future, a possible openness to a secret that differs from the one that has marked metaphysics with the structure of Western thought. The only question is with what other opinion can the future be freed from the nihilism with which life sinks into formlessness and chaos? The pre-Socratic philosopher who first used the word *logos*, Heraclitus, may have had a solution to the spiritual discomfort we face day and night. In the seventh fragment, he said this:

> If you do not hope, you will not win that which is not expected, since it is unattainable and inaccessible. (Heraclitus, 2019)

To think means to approach what the unexpected event allows.

References

Agamben, G. (2019). *Creation and anarchy: The work of art and the religion of capitalism*. Stanford University Press.

Blanchot, M. (2015). *Literary space* (V. Šeput, Trans.). Litteris.

Deleuze, G. (2011). *Différence et répétition*. PUF.

Heidegger, M. (1975). *"The Anaximander fragment" in early Greek thinking: The dawn of western philosophy* (D. F. Krell & F. A. Capuzzi, Trans.). Harper.

Heidegger, M. (1998). *Einführung in die Metaphysik* (6th ed.). M. Niemeyer.

Heidegger, M. (2002). *Was heißt Denken?* (Vol. GA, Bd. 8). V. Klostermann.

Heidegger, M. (2003a). "Die Zeit des Weltbildes" (1938). In *Holzwege* (8th ed.). F.-W. von Herrmann.

Heidegger, M. (2003b). "Der Spruch des Anaximander" (1946). In *Holzwege* (8th ed.). F.-W. von Herrmann.

Heidegger, M. (2005). *Von Anfang* (Vol. GA. 70). V. Klostermann.

Heidegger, M. (2007). Das Ende der Philosophie und die Aufgabe des Denkens. In F.-W. von Herrmann (Ed.), *Zur Sache des Denkens (1962–1964)* (Vol. GA. 14). V. Klostermann.

Heidegger, M. (2012). *Der Anfang der abendländischen Philosophie: Auslegung des Anaximander und des Parmenides* (Vol. GA, Bd. 35). V. Klostermann.

Heraclitus. (2019). *Fragments. https://pressbooks.pub/sapientia/chapter/heraclitus-fragments/*. Retrieved October 11th 2021.

Nietzsche, F. W. (1996). *Homer's contest. Nitzscheana 5*. North American Nietzsche Society.

Paić, Ž. (2018). *Tehnosphere. Vol. II, The "Black Box" of metaphysics: Cybernetics and the absolute time of the machine*. Sandorf and Mizantrop.

Piementel, D. (2019). *Heidegger with Derrida: Being written*. Palgrave Macmillan.

Schürmann, R. (2003). *Broken hegemonies*. Indiana University Press.

Sloterdijk, P. (1986). *Der Denker auf der Bühne: Nietzsches Materialismus*. Suhrkamp.

CHAPTER 4

The Body and the Technosphere: Beyond Phenomenology and Its Conceptual Matrix

4.1 Introduction

No one described the destabilization of metaphysics better than Nietzsche. In *Thus Spoke Zarathustra,* he claimed the imperative of radical change in metaphysics and its values by eternal recurrence (Nietzsche, 1988). Yes, returning what was given out of pure generosity is now shown to be the task of establishing "new values". It is not something only symbolically related to man and his habits of thought. It means a change in the very way of living. When the beyond loses its significance, the immanence reigns supreme. Furthermore, instead of the subject, the term object gains new legitimacy in determining reality. What was despised for centuries as transitory and in the service of the spirit—*the body*—suddenly takes on the functions of a thinking object.

The problem faced by thinking in the twentieth century stemmed from the body's entry into an existential abyss, as reflected in diverse physiological and psychological theories. Among them, above all, psychoanalysis tried to penetrate the dark zones of the unconscious, starting from the individual subject as the guardian of language. And thus, the question of the objectivity of the object focused on the sublime in the fetishism of things. This path leads from illumination to fascination. The body turned out to be an ontologically "empty center of power". The writing of signs into its signifying void could begin only after liberation from the rule of the *logos*.

Ž. Paić, *The Technoscientific Turn of Philosophy*,
https://doi.org/10.1007/978-3-032-21081-4_4

How is it possible to think about what connects this emptiness with its manifestations, ranging from desires and the will to power to cognitive processes of creating a complex reality, all the way to the relationship with the central concept and problem of contemporary technology, such as technology in the form of the technosphere? In what follows, this path from metaphysical to cybernetic difference, or from ontology as a phenomenology of the body, especially in the work of Maurice Merleau-Ponty, to the open space of *techno-genesis* of objects is considered. The starting point is that the body must be understood historically and epochally as a machine in its two modes—analogue and digital. Transformations of Being (*nature*) belong to the former case, and transformations of events (technosphere) belong to the latter case. However, the machine that unrolls the body at an irreducible speed of transformation means that neither the logic of machines nor the structure of the organism meets the definition of a machine and a body in general anymore.

We must therefore immediately develop a language for new phenomena. Before that, it is necessary to free thinking from the habit of vainly seeking meaning in the idea of cause and purpose, and instead to understand how the technosphere peers into singularity, reaching the limits of the visible and the invisible worlds of objects. For its fundamental principles, only visualization is sufficient to know what is happening as a movement in space and time, and to recognize its irreversibility. Nothing is repeated without a "new" way in which the original and the simulation, the stable and the unstable, are creatively mixed. Opening the problem of techno-aesthetics of autonomous objects means reconsidering the role of the body in the creative process of *techno-genesis*. But it is also possible that it will disappear in the process of cybernetic disembodiment. The design of the body, therefore, marks the transition from the aestheticization of the world to a ready-made object to the dizzying techno-aesthetic construction of a "new" life. Marcel Duchamp and all theories of design as applied art have nothing more to say on the subject. When life becomes technologized to the point where autonomous objects exist, everything becomes possible and real.

4.2 The Phantom Limb: Maurice Merleau-Ponty and Borders of the Phenomenology of the Body

The keyword for the painter's existence, "Being in the world", originates in the verb related to real and phantom limbs, serving as the central organ of the artist's physical-cognitive engagement in the situation. *To handle* something does not mean to carry out the hidden will of the transcendental subject in the sense of the initiator of the action. Then the spirit would "manage" the body at its discretion. At the same time, subjectivism would be written into the blind destiny of an organ with a special place in the determination of human "being". We do not mean the symbolic or metaphysical meaning of the hand that governs human life in the way of directing it into the socio-political sphere of command, nor of conducting a symphony orchestra or, on the other hand, pointing to the act of faith and grace with which in Michelangelo's allegory of the creation of the world on the walls The Sistine Chapel in the Church of St. Peter's in the Vatican signifies the touch of Human and God. To handle refers to dealing with something that is already serving another purpose. In this respect, the hand has a double meaning. It is a means of action and an instrument of direction towards the goal of action.

As part of the symbolic-volitional activity of humans, it can even be said that without a hand, humans cannot be an "operative" system of functional action. It is therefore not surprising that the problem of the so-called phantom limb (organ), i.e., the phantom hand as a prosthetic or technical replacement for what has been mutilated or taken away, reveals the key difference between ontology and cybernetics. The first focuses on nature as a necessity for bodily functioning, as an "innate" mode of connection between instincts and sensibility, and on the soul/spiritual manifestations of "Being in the world". The second, by contrast, concerning the construction of the artificial body (robotics and engineering), is oriented towards the freedom to think and act in new ways. Formally speaking, it has this advantage because it comes from the work of "artificial intelligence".

The paradox is that freedom and contingency are conditions for the possibility of the cybernetic system managing the world as an open field of options. At the same time, aporia is reflected in the fact that "nature" is determined by the non-freedom inherent in the facticity of bodily existence. Man and other beings cannot choose this voluntarily. Acting according to the principles of freedom begins where "Being in the world" is always limited by fate and bounded in space. Hence, the "phantom limb" is neither a rhinoceros horn nor a tortoise shell, but a technically or mechanically created so-called third hand that manages the processes of exchange of matter, energy, and information from two worlds: (1) natural or analogue and (2) technological or digital. For the first world, language becomes a condition for knowing the world. For the latter world, it appears necessary to learn the rules of visual semiotics, because the techno-image lies at the "essence" of non-human communication. Merleau-Ponty says about it in *Phenomenology of Perception:*

> A hybrid theory of the phantom limb, which found a place for both sets of conditions, may, then, be valid as a statement of the known facts, but it is fundamentally obscure. The phantom limb is not the mere outcome of objective causality; no more is it a cogitatio. It could be a mixture of the two only if we could find a means of linking the 'psychic' and the 'physiological', the 'for-itself' and the 'in-itself', to each other to form an articulate whole, and to contrive some meeting-point for them: if the third person processes and the personal acts could be integrated into a common middle term. (Merleau-Ponty, 1958, pp. 88-89)

Let us examine the reasoning that follows from the distinction between reflex actions in animals and humans. We will see that this understanding, along with Lacan as the primary representative of the new psychoanalysis, does not differ significantly from Heidegger's approach to the relationship of stone, animal, and man to Being. While, namely, existential phenomenology attributes to the human body the possibility of spontaneity and reflex action only if it is engaged in "Being in the world" situations, and this means that practical knowledge takes precedence over theoretical mere fact, an animal cannot relate to Being except instinctively and reflexively, through action. Admittedly, Merleau-Ponty will not say that because of this, the animal has no world or that the world is less valuable to it than the human world. However, it will also not contradict the basic assumptions of philosophical anthropology. According to them, some biological and cognitive evolution contributed to the hand and brain directing all further operations of thought.

All this testifies that the body cannot be absolutized by establishing human existence in spatial terms through the immanent transcendence of the openness of Being in general. Consideration of the so-called phantom limbs primarily serves a cognitive-theoretical function, focusing on the essence of metaphysics. In the "idealization" of the permanence of Being and the perfect order in which diverse beings live in harmony and conflict, metaphysics never saw the body as lacking "in-Being" as such, with dissymmetry, disharmony, and the deconstruction of the world. Therefore, its language cannot accommodate hybrid systems of difference, and in this chaos of contingency, it cannot admit what is simple, painful, and imperfect. What? Those people are mortal and prone to pleasures, sick and perverse, neither angels nor demons, but beings of physical existence with the aspiration to achieve immortality by moving into a *posthuman condition*. With the help of the body, man is aware of the world. This realization is the root cause of human irrationality. Thus, existence becomes a condemnation of freedom and meaning.

What does the statement about the "ambivalent presence of the hand" mean? Let's not forget that Merleau-Ponty published The *Phenomenology of Perception* in 1945. In other books written in the 1950s and 1960s, we rarely encounter examples from already established fields of computer science and cybernetics. The same applies to Jacques Derrida's major work, *Of Grammatology*, from 1967 (Derrida, 1967). It explicitly employs the concepts of cybernetics and semiology, including information, code, program, feedback, signifier, and signified. However, the phenomenology of the body has the task, first, of establishing the existential organization of reality outside of consciousness as intersubjectivity.

Although Merleau-Ponty dares to assign Husserl the position of the leading thinker of the path towards the existential turn in contemporary philosophy, which is in opposition to Sartre's propositions from his phenomenological ontology in the writing *Being and Nothingness (L'Être et le Néant)* from 1943, (Sartre, 1943) with noting that Sartre is the first to decisively place the problem of body and existence on the horizon of his reflection, it seems that it is still much more important to notice his connection/relationship with Heidegger's concept of "Being-in-the-world" (*In-der-Welt-Sein*) (Heidegger, 1977). The ambivalence of the body cannot be understood without the massive assumption that the body is an existential projection of the meaning of "Being in the world". By itself, it has no other meaning than the physical, physiological, and psychological structure of sensibility. What distinguishes a Human from an animal cannot be the mere presence of his body. At its centre lies the existential relationship of a man who suffers, feels, wants, loves, creates, and thinks only because he is an experiential being of physicality.

The "phantom limb" in the phenomenology of the body cannot be considered a supplement/replacement of Being in the form of nature, to use the term of the early Derrida in *Of Grammatology*. The reason is that its provision should be hybrid. On the one hand, it concerns the hand as a natural human organ; on the other, it concerns a foreign body that mimics the actions of a real hand. The problem, of course, is that it is not art in the sense of work, such as an installation displayed in a museum. Mimetic action no longer refers to imitating nature as such, but to the techno-poietic system of "operational thinking". This is a highly complex relationship among necessity and freedom, reflexive actions, and volitional-cognitive activity. Since humans do not use their hands solely for everyday purposes in their mechanical existence, it is not possible to mechanically replicate the model of nature or the analogue world. Many examples from medical practice, such as amputation and augmentation of the "third hand", reveal subtle relationships within the sensory spectrum of pain and suffering, joy and elation, mourning and sadness, and feelings of pride and self-recognition. A man, faced with the necessity of accepting a mechanical prosthesis for mere survival, becomes someone else.

However, this does not mean complete personality change. He only clearly perceives that his body denotes the medial area of the permeation of life as a connection/relationship between nature and artificiality. Of course, he sometimes feels pain as if it were the memory of the original living hand that is no longer there, and instead of it, all the operations are now performed by the "third hand". In the exhaustive analysis of this condition in which the "phantom hand" operates, Merleau-Ponty provided the basis for an almost identical procedure of refined analysis of the acceptance of the transplanted heart as a "foreigner" and a "living machine" in the essay "Intrus" by Jean-Luc Nancy, included in his book *Corpus* (Nancy, 2008a, pp. 161–170).

Is it possible to generalize the experience of the singular individuality of a person who, thanks to a mechanical prosthesis "on" his body or the installation of an apparatus "in" his body, becomes someone with a different experience of the world? The body becomes the primary experience of *my* body. Heidegger most clearly articulated the critique of Kant and Husserl on this in *Being and Time*, which discusses the structure of Being there (*Dasein*) as a Being-in-the-world, in the mode of my always assuming the determination to understand Being in general. *Mine* is not about *anything* vs. the *world*. For something to be appropriated and marked as *mine*, the content of consciousness "about" the world must first be reduced to the openness of perspectives.

Only in *my* world, even that monstrous "phantom hand" can be called *mine* on the condition that it belongs to the structure of the autonomous will-feeling of some Self. However, it is no longer its extension with respect to the Cartesian body. Now the "third hand" with its technically produced "will" belongs to *my* existential space of thought and action. Even more precisely, *my self-label* must be expanded to encompass not only the surrounding world *(Umwelt, environment, milieu)* but also the technical landscape of Being. If, on the other hand, we ignore, for methodological reasons, that spatiality can no longer be expressed in the technical landscape of corporeality by oppositions of external and internal, we are left to see what this "phantom hand" truly means in a new meaning beyond metaphysics and its derived concepts and categories.

Merleau-Ponty explicitly claims that "the phantom limb is not a simple effect of objective causality, but it is no longer a *cogitatio either*" (Merleau-Ponty, 1945, p. 110). The main reason for the introduction of this term, which does not seem phenomenologically correct, is that it is more reminiscent of the psychoanalytic language of the difference between the phantasmatic and the real, so it would suit Lacan perhaps even more than it seems at first glance, to show how the body appears as an object. However, it should be clearly emphasized that this is not a classic contradiction arising from the metaphysics of subjectivity. Namely, for metaphysics, the object should always be a construction of the subject. By contrast, the tradition of French materialism, exemplified by La Mettrie and Holbach, was introduced into philosophical thought, in contrast to Kant's concept of the thing-in-itself (*Ding-an-sich*), which posits a mechanistic conception of nature. In this context, the body's thought appears for the first time.

However, in contrast to the Cartesian dualism of mind and body substance, there is now a solution that has a closed circle of matter's action without the first mover and the last purpose. The objectification of the body does not mean its transformation into an inanimate object, like a stone-like "thing", or, on the other hand, transitioning into the form of mechanical existence of a prosthesis without the participation of the organic. What is an object for existential phenomenology? Nothing but the product of external and internal relations between subjects in the perception of the world. Husserl would add here the keyword for the solution to the problem of the relationship between the Self and the community: intersubjectivity. However, when the body appears as an object, then it is always something other than the subject, that irreducible area from which stimuli for action come, because the object is not a mechanical toy without a "soul". Its appearance requires reflection on the conditions of possibility under which the body is de-subjectivized and becomes more extended than the psyche, as Freud said in a posthumous note.

The "phantom hand" is therefore not moved by God or another human being. The desire to objectify the subject turns out to be a decisive factor. Thus, the perception of the external world becomes a problem of determining what is real. In the classical projection, it was a site of synthesis between consciousness and Being, manifesting in two modes of appearance in modern philosophy: idealist and materialist. To be an object, however, Merleau-Ponty means to leave the Scylla of "objectivism" and the Charybdis of "subjectivism" in the footsteps of Husserl. This means, concretely, addressing the problem of a phenomenon that no longer has anything to do with the metaphysics of nature or with the idea of Being as constancy in change.

What makes the fundamental criterion for separating animals and humans for existential phenomenology? We will not say that an animal lacks a body. However, we will not equate cases of replacing limbs with those of insects or antlers with those of deer by analogy with a "phantom hand". The phantom in the hand is nothing but the presence of life as a form in the technical event of state transformation. The hand can be formally replaced indefinitely. The infinite sequence resembles a digital copy of an image. Namely, there is no original here. However, one must not lose sight of the fact that the "third hand" does not take the place of the "first" as a mere thing without a "soul". The copies are, admittedly, indistinct in the same way. But that is not the singularity of a living organism. Its fateful expulsion into the world is that the organism shows as irreplaceably sensuous in suffering and pleasures.

Each "phantom hand" does not sit on living flesh as a replaceable organ according to the model of analogical nature. Instead, we face the uncertainty and contingency of events. What will happen when a foreign body, that uncanny otherness of a technical organism, is received or accepted cannot be predicted in advance. This is precisely the essence of the body's existential organization. It becomes the indisputable living singularity of the transformation of organs or limbs. The totality is not superior to the parts, as in Hegel. Between these exist the logic of singular reproduction, which becomes different from the logic of producing difference. An animal can reproduce only unconsciously by employing the replacement of limbs, and Human leads existence to the highest level of what Dante calls the "new life" (*la vita nuova*).

The search for the "new" utilizing substitution does not mean that the technique might be understood as a mere mechanical means for other purposes. In the phenomenological and psychoanalytical search for a solution to the problem of creative "human nature"—and this is true for Merleau-Ponty and Lacan—there is no point of transition to the essence of technology. However, places of mediation, informed by the newer results of cybernetics and semiology in the 1960s, were frequent in their works (Paić, 2019). Why is that so? The answer that seems to be acceptable is that the concept of existence as an essential new "essence" of man and desire/unconscious as a structure of corporeality in the world does not reach what the most critical thinkers of technology in the twentieth century, Heidegger and Simondon, credibly opened as the main problem of modernity. How, namely, to preserve the experience of different thinking against the logic of technoscience, without at the same time falling into the fold of the overplayed metaphysical scheme of history about humanity as an authentic Being and inhumanity as the vulgar existence of a technical object?

Let's return to that strange and irreducibly ambivalent "phantom hand" in a moment. Merleau-Ponty describes the experience of the mutilated body of the subject with psychoanalytic language. It is, therefore, a "traumatic experience". "A *certain anxiety*" arises from the realization of its inexplicability within categories of vitalism and the organic attachment of the body to the Earth (Merleau-Ponty, 1945, p. 112). Handling objects is possible only under the conditions of some primary contact between the living and the non-living. Moreover, the phenomenon of touch, which Nancy insists on philosophically, shows us that in the case of stroking an "Obsidian head" or a marble statue, there is still some excess of the desire to objectify. The coldness of the statue and its perfect indifference, stemming from its inability to reciprocate touch with a sensory reaction, cannot serve as a model for understanding the technical apparatus in the living body.

Touching objects or living beings is not the same. The reason is that the technical replaceability of organs in the "other way of life" requires the hybrid creation of events. Feelings and experiences of something that cannot be clearly described are such uncanny events. The arrival of a mechanical machine in people's everyday lives causes discomfort, disbelief, and astonishment. This lasts only briefly because the technical existence of machines, robots, and cyborgs, as Gilbert Simondon put it, is domesticated as a "foreign body" within the socio-political environment of humans. The process of accepting the inhuman is no longer such a traumatic experience if the core values of the community are not questioned.

Let's remember that today, the relationship between the achievements of technoscience in medicine, such as reproductive stem cell cloning, is tolerated by religious communities. But only to the extent of the distance between the so-called untouchability of Natural Law (God) and human intervention in the biological default of the organism. When that limit is crossed, serious disputes arise. In this regard, bioethical norms are constantly changing. For the most part, they depend on the level of the value scale concerning the body problem in modern society. This is additionally ethically challenging. The reason lies in its revealing the impotence of traditional metaphysics and the religious-ethical doctrines built upon it before the emergence of transhumanism and posthumanism. It is enough to extract the main argument for prenatal selection and reproductive cloning: the desire for healthy offspring! (Paić, 2011, pp. 65–117).

However, it is not only faith as the foundation of religion that is brought into question here. Keeping the memory of the primordial nature of Being and the changes that do not call it into question means defending what has long been indefensible. In his phenomenology of the body, Merleau-Ponty started from the assumption that imperfect beings inhabit the world. Moreover, these beings, especially men among them, are the least understandable in their mutual relations, starting from what one thinks of the other and vice versa. The first fact that we encounter is the view of the Other's body. In contemporary French philosophy, apart from Sartre and Merleau-Ponty, this idea was most strikingly articulated in the language of phenomenology by Emmanuel Lévinas. His turn from ontology to ethics served to research the world according to the measure of human impotence and the freedom of unconditional commitment to the Other (Lévinas, 1990). In all three cases, the body is shown to be the main ontological problem, for the simple reason that it concerns the meeting between objects in space, the contact between beings, and the existential restlessness and discomfort that Being "is" and that this happens in unpredictable, unexpected relationships. If beings are imperfect, then the fundamental impulse of their meeting becomes an attempt to free the body from the stigma of the ideal world and the form in which that world mystifies itself to the extreme limits of the sustainability of the order of concepts on which its metaphysics rests.

Body mutilation in the context of a "healthy society" arises as an excessive phenomenon. The loss of body integrity through mutilation also causes discomfort in the observer. This kind of shyness often leads to pathetic compassion for people with disabilities. However, the panicked need for healing and normalization paradoxically humanizes the world's technical character. All this occurs only under the condition of a transition to a posthuman condition. Traces of the "phantom limb"are visible in the talk. For Merleau-Ponty, it was a necessary step towards a different determination of the meaning of existence. Without a body, everything seems just the appearance of a Being, a deceptive sublimity without nature, an insight into the blueness of the sky in the night of the end of history. Things are, therefore, upside down. Their perspective is visible only from an unusual angle.

The body, therefore, becomes a scandalous act of openness without any shred of theodicy according to which salvation comes after the end of the body and relates to the soul of man. We can safely claim that the two most significant "big narrative" theories in the humanities about the body are phenomenology and psychoanalysis. Both address the effects of modern natural and technical sciences on the body's experience across all aspects, from medicine to engineering. However, both "big narratives" are based on concepts that belong to Western metaphysics, albeit in their descent from the throne of ideas in the form of inverted Platonism and Christianity. Nietzsche expressed it best. In the intercessions of the will to power as an eternal recurrence to the fundamental concept of "life", he opened the space for the act of radical deconstruction of Being. And indeed, the body as an object can never be understood otherwise than as the opposite of "life". This is why the desire for immortality primarily becomes a desire to prolong human physical existence.

In analysing Merleau-Ponty's statements from *Phenomenology of Perception* and other writings published in the 1960s, it becomes evident that the body is restored to its dignity in thought only when that of spatiality replaces the primacy of temporality. This is not only a significant difference compared to Heidegger and his intended notion of the "second beginning", which begins with the mission of Being as an event

(*Ereignis*). Hence, spatiality becomes the authoritative way of techno-genesis of autonomous objects. In other words, Merleau-Ponty represents the beginning of the thought of recognizing the eccentric and extravagant bodies of the human-non-human. By determining Being-in-the-world through the existential organization of the body in practical engagement, it became possible to abandon the Cartesian relics of body thought. The body cannot be just an existential being, as Jean-Luc Nancy says in *Corpus*. From the extensibility of matter, the supply of energy, and the deliverability of information, it cannot be reduced to what already is, that is, to be in the permanence of changes.

My position is that, following Merleau-Ponty, the body should be understood as an elementary existential event. The meaning of that event cannot be predetermined, nor does it appear at the end as a hidden secret of Being. Instead, it would be necessary to consider the body in its ultimate potential for transformation. Like Kafka's character, Gregor Samsa, who is transformed into an insect in a dream, the body also opens as an event beyond any previous ontology. Jean-Luc Nancy is correct in this regard when he asserts in the *Corpus* that the body has yet to be thought of ontologically (Nancy, 2008b, p. 15).

The only problem is that for such a different thought, ontology can no longer be authoritative. This may also be true for many other cases, such as image ontology. The reason lies simply in the fact that the vertical and hierarchical model of understanding the world, with God as the central substance and Being as the leading concept, no longer corresponds to the procedural process of creating many virtual worlds. Instead of phenomenology, which could place the problem of the "phantom limb"in the centre, still, as a continuation of mechanical technique by other means, because only in the 1960s will the first transplant of a human organ such as the heart take place, the body in the cybernetic way of thinking can no longer be determined, even positively nor negative. It is neither an extended substance (*res extensa*) nor is it a function of some phantom human intelligence that feels even the finest vibrations of the Earth. If all that body is not, then what "is"? Nothing. Yes, you heard right—nothing. It does not exist as an entity in itself. It is also not conceivable as a thing-in-itself (*Ding-für-sich*).

No Enlightenment epic about the process of developing a higher level of consciousness in the body as a neurocognitive network of plasticity gives the last answer to the question about what, after Wittgenstein, is called "language games"(*Sprachspiele, know-how*) in the philosophy of language. With this, we have already indirectly indicated the solution to the problem. For the body to be thought of as the initiator of the *transformation of events,* it is no longer possible to understand this starting from any ontology. Its universal application to diverse areas of Being has passed. All the so-called regional ontologies that Edmund Husserl was still talking about are now melting away in the flourishing of a multitude of aesthetics. However, this is not proof of the absolute predominance of philosophical thinking in the age of the *technosphere*; rather, it indicates the complete fragmentation of knowledge about the worlds of pure construction. Instead, it is necessary to start with the initial assumption that, as an insect replaces its organs, the "phantom hand" can be replicated through a technical process of event transformation, which is already performed today in medicine using a 3D printer.

What does this change mean in the understanding of the body? Almost everything that was once considered normal. The impossibility has become a pragmatic possibility for transformation in the functioning of the body as a technosphere, transcending the distinction between living and inanimate. If, for example, one wants to improve a person's ability to remember in a complex situation that requires a high level of intelligence, as is the case in space flights today, the solution lies in improving the operation of the "artificial intelligence"device, and not in the birth of a potential genius. A technical understanding of the body eliminates any reliance on pure essences, a concept phenomenology previously addressed. Instead, we are dealing with a pure "uploading" of cybernetically created protocols of the body-as-machine.

The only aspect of Merleau-Ponty's existential phenomenology that remains relevant in the age of the technosphere is the problem of determining the existence of the uncanny inhuman that conceals the idea of the "phantom limb". What kind of existence is it if its essence should be reduced to calculation, planning, and construction, and no longer to incalculability, singularity, and unpredictability? Does it have something more than the horizontal arrangement of events without foundation in the idea of a creative original, which appears under the name of simulacrum, and reproduction as a condition for the possibility of the emergence of new machines of contingency? We can answer these questions only when we establish the essential difference between emergence and *techno-genesis*, the transformation of Being as becoming (*Werden, devenir*), and the transformation of a condition as an event (*Ereignis, événement*).

4.3 Contingency Machines

The classical philosophy of technology, such as Ernst Kapp's, must have been derived from the book *Basic features of the philosophy of technology: A contribution to the history of the emergence of culture from new points of view* (*Grundlinien einer Philosophie der Technik. Zur Entstehungsgeschichte der Kultur aus neuen Gesichtpunkten*) from 1877. The motto seems decisive. Kapp cites the thought of Edmund Reitlinger, who says

> That the whole of human history, if the truth is taken, can ultimately be considered the history of finding a better tool. (Kapp, 1877)

This reduces the essence of historical development to technical inventions, focusing on the improvement of things that serve human purposes. However, if we purify this "philosophy of technique" from the classical ailments of modern dogma, such as the law of causality and the rest of natural or rational theology, in the notion of the purposefulness of history, what do we get? Nothing but the problem of "progress" and "development" of automata, devices, and things that belong to some indifferent world of pure objectivity.

After all, even the word "tool", which derives from the Greek "organon", encompasses the meanings of a logical system, a method of using it, and a way of managing it, as the manipulation of an object beyond human organic purposefulness, which points to this self-sufficiency. It can even be said that technique is reduced in everyday dealings with life to that existential space. Newton defined the space in which two-dimensional objects rest as a "collector" or container. Things are, therefore, essentially "not" because their Being as technical bodies cannot be derived from the phenomenological concept of existence as proposed by Maurice Merleau-Ponty for the modern idea of freedom. In addition, the problem of technique does not lie only in the vagueness of this self-sufficient indifference that we observe in the graveyard of old machines from the mechanical era.

Therein lies the paradox of the object's technical existence. The faster the obsolescence, the more schizophrenic the need for new ones. Without this paradox, the object has no reason to exist, and the machine remains an empty flywheel of motion. What we call a machine is not a mechanical device in motion. On the contrary, the machine includes *organon*, *téchne,* and *poiesis*. While the machine is reduced to the inhuman in the sense of insurmountable opposition to the human Being, here we encounter the trinity of *management, performance,* and *production*. In this way, it can be said that in the cybernetic system of the technosphere there is a synthesis of the management mechanism and control over the processes of producing new things (objects and data networks), the performative concept of knowledge as a pragmatic use of language in the form of a visualized concept, and last and not least, infinite production "forms of life". They should incorporate elements of nature into the analogue world and artificial life into the digital world (Rieger, 2003, pp. 315–326).

What does self-sufficiency mean concerning the ontological status of technical devices and machines? It would be a mistake to think here that the technique in the mechanical way of working tools has some special independence of its own. A thing as an object of the subject's reflected experience cannot have the autonomy that the mind has, which, as Kant says, is its legislator. However, in the theological sense of the word, only God is assigned this lofty idea of freedom because it is not limited by anything external. Hence, the origin of modern theories of political sovereignty is in the theological science of God's unlimited power. Nevertheless, a kind of limited autonomy, which is a higher form of self-sufficiency than what the Greeks called *autarky*, belongs to the field of technology in the sense of reshaping nature.

Matters change significantly when modern technology based on automation is introduced. When machines operate independently to perform complex tasks that require a higher level of cognitive and volitional capacity than humans, and when humans are relegated to supervising the factory process, we are already on the path to machine self-sufficiency. Gilbert Simondon calls it "the second order of cybernetics". However, the historical development of techniques from the modern era to modern technology defined the period from the nineteenth-century industrial society to the end of the 1960s. Sovereignty and self-sufficiency belong only to the "third order of cybernetics". Here, information precedes matter and energy, and the management or control system is based on the idea of a feedback *loop*.

This circular irreversibility characterizes the process of liberating the technosphere from all mediations and medial reflections "about" the world and the Human. Instead, we are faced with a machine that thinks for itself by producing events such as state transformations, rather than mere objects, through the technique of replication and cloning. In this sense, the technosphere is truly self-sufficient. It cannot be compared to the graveyard of industrial technology simply because it is on the path of absolute dematerialization.

Contemporary quantum computers mark the beginning of the second digital revolution, for which Merleau-Ponty's concept of the "phantom limb"holds little relevance, except perhaps in evoking the era when the body was still understood in terms of the absolute spontaneity of human will. However, today, the very concept of free will in the traditional metaphysical sense is called into question. Neurocognitivists and enactivists no longer speak that discourse. The brain "thinks" pragmatically. In its plasticity, it reacts to events by going beyond the arguments of physiologists and psychologists, those who attribute everything to reflex drives and those who bet on the magnificent and irreducible space of human action, encompassing voluntary decisions (Sturma, 2013). In this almost inexplicably childish "fascination with the brain", as Jan Slaby points out, we are witnessing the natural sciences in the guise of new objectivism supported by powerful visualization techniques (SCAN) taking over the once unconquerable territory of philosophy and spiritual sciences (Slaby, 2014, pp. 211–221).

Self-sufficiency, autonomy, sovereignty, and the absolute rule of managing the system and its surroundings become fascinating ways of unfolding "life" in highly developed contemporary societies. Why is that so? It seems evident that there is something uncanny and simultaneously amazing in the "essence" of the Technosphere, as it encompasses the concept of new information and communication technology as a "thing that thinks" and an "apparatus/device that is aesthetically seductive". Ambivalence arises from its indefinite ambiguity. It is both a *thing* and a *creature* in the sense of a cybernetic virtual avatar; its calculated images are not perceived as images of nature, but as the creation of a new reality, with a fundamental shift in the "essence" of the image. It does not depict or represent an objective world. Technosphere calculates, plans, and constructs new worlds.

Suppose we return to the problem of determining the reason why the phenomenology of the body no longer has the possibility of insight into what is happening with the techno-poietic way of transforming the condition, by which bodies can be reshaped not only by transplanting organs and replacing them with other organs, but also with animal organs. In that case, we immediately arrive at the central issue of this controversy. That is *history*. That is why Hans Blumenberg is right when, in his analysis of the relationship between phenomenology and technique in Husserl's late writings, he shows that for him, history is.

> *Nothing but a living movement that which is common and that which permeates the original conception and sedimentation of meaning.* (Blumenberg, 2015, p. 175)

The disappearance of living history from the scene must be replaced. Therefore, the new cybernetic physicality lies in a constant state of transformation. The stability of the system results from its change. Everything valid for the technologization of the language describing these bodies as becoming disembodied in the *robot-cyborg-android trinity* is even more valid for the technologization of the image. Why do we discuss various relationships in the construction of "artificial life" (A-life) in relation to language and image? The reason lies in the fact that language is telling and therefore has the communicative potential of symbolic exchange in the common Being of man. Artificial languages can only be algorithmic in their use for visualizing concepts. In other words, their function becomes instrumental, and that is why the process of technologization always mediates it. In contrast, a technically produced image is completely aesthetically autonomous. Without reference to anything in the given world, the information refers only to itself and to other images. Hence, the process of technologizing the image focuses on objects in the space of virtual interaction. Language continues to be used to describe things and phenomena.

The picture shows only what occurs during the continuous transformation of events. Language, therefore, belongs to the realm of ontological difference, while the image is derived from the cybernetic difference. To the first, history appears as transcendentally a priori, and to the second as immanence in a movement without beginning and end. It necessarily follows that we no longer live in "worlds of life", but in "forms of life". Original or immediate life is reduced to the structures of "bare life", and moments of unique happiness are almost rare. Around us are numerous digital platforms. They multiply under conditions that enable new physical or visual communication. Gone are the days of unexpected encounters and uncertainty. Now, the only question is how contingent machines produce desires and resistance in the world, because the structure of life is not created "naturally" or "historically". All this is far behind us, like a pale shadow of things in the accelerated "aesthetics of disappearance" (Virilio, 1991).

Having a body today does not mean being condemned to a singular conception of the world. For Merleau-Ponty, the phenomenology of the body was an onto-pathology of living corporeality, leading to a "phantom limb"experience. In contrast, one could speak of a historical way of existence in which the body engages in various situations. Namely, it was only possible because the body had an original "flaw": an ontological defect in that, unlike insects, for example, it could not *auto(re)generate*. The rapidly developing technology, based on the reproductive matrix of copying originals, was initially limited to mechanical and semi-automatic devices. Let's focus on horror films from the experimental avant-garde phase of the 1920s. We will understand that the dismemberment of organs and mutilation of the body tends to be visually supported precisely by the feeling of disgust (abjection) towards the monstrous event of the destruction of the integrity of the body. In analogy with the "phantom limb" which pathologically questions "Being-in-the-world", the film staging was based on the excess of "phantom images".

The cut on the body—the cutting of vital organs—functions as a perceptual shock in the films of Salvador Dalí and Luis Buñuel's *Un Chien Andalou* and Germaine Dulac's *The Seashell and the Clergyman*, directing the viewer's attention to what lies beyond the shock itself. The sublimity of the experience of the thing itself, with which the gaze enters the space of chills, connects disgust and monstrosity. There is no better term for it than the German word *Unheimlichkeit*. It is about the outrageous fear and admiration of what is both foreign and close to man. The body that disappears in the self-sufficient, autonomous unfolding of the technosphere ends this effect of *Unheimlichkeit*. Moreover, by its suspension and neutralization as a constructed object of self-staging in virtual space, the body becomes a replaceable singularity of the case. Nothing seems impossible anymore, and everything becomes a performative event—from plastic surgery to the birth of a monster—as in Dalí's painting *The Geopolitical Child Watching the Birth of the New Man* in 1943.

Today's research on the human body, concerning technoscientific constructions, reveals how much "operational thinking" is inscribed in the procedures of "embodiment", going beyond the prevailing attitude that the mental substance determines and orders the body's actions, marking the end of modern subjectivism. The discussion of the "objectification of the will" as an offshoot of metaphysics, as presented by Schopenhauer, testifies to the understanding of knowledge in terms of the mind-body distinction. Therefore, the body could only be understood as an object of knowledge or an intention of free will in the sense of the action of *logos, spirit, and mind*. Incarnation is either the descent of the mind into the body from the heights of transcendence (philosophy) or, on the other hand, the shaping of man in the image and likeness of God (theology).

In both cases, it is understood as an object, although the Greeks, unlike the Christian concept of resurrection, regard the dead body merely as a corpse. As we have already demonstrated, in phenomenology, the body serves as a subjectivized means of existentially engaging with the world (Husserl and Merleau-Ponty). This marks a significant departure from Cartesian "functionalism". Today's attempts to understand the body's complexity integrate neuroscience and cognitive science. The upheaval occurred precisely because research on machines and artificial intelligence showed that humans cannot be unambiguously classified, neither with respect to transformations of Being nor concerning transformations of events. Its "nature" is that which mediates between the two shores of the world-historical existence of the technical world. Being situated between "nature" and "technology" enables the body to merge with and permeate something that transcends duality.

Is this age a sign of the absolute rule of the flesh, or is it just an illusion? The answer seems to be derived from the logic of contemporary action, which is either too broad or too narrow. Yes, the body appears everywhere in transformations of Being. Contemporary art, for example, is defined through a performative-conceptual turn. The same applies to efforts in the interdisciplinary field of transhumanism. Here, by contrast, research focuses on enhancing the physical structure of humans within the technosphere. The fascination with physicality stems from an interest in the image encoded in digital form. *Instagram, Twitter, Facebook,* and other social networks, in addition to showing mass idolatry of sexuality and the body, also testify to the narcissism of our sophisticated technical era. However, everything is directed towards disembodiment, the movement towards the Omega point of the universe.

The ambivalence of the image as a body and the body as an image permeates all human activities because, in the age of the technosphere, the body cannot be merely a "thing" of philosophical-theological "embodiment". Instead of the mystery of the entry of spirit and soul into the body as flesh, the fundamental issue appears as how it happens (*quoddittas*), "what" connects nature and technology. In other words, extending the domain of the "phantom limb" into the world of life demonstrates that living in a network of "phantom images" requires processes of aesthetic "embedding" *the implant* into the body. At the same time, their structure lies between the living and the non-living. To be aware of one's own body today means to move from the existential drama of Being to the pure indifference of *body design*. This turn introduces us to a space in which powerful machines of contingency operate noiselessly and glow with fluorescent light without stopping. Spirit and soul have passed their own forever. The only thing left for the body is a "bright future". But can we still call "what" we encounter in our daily dealings with information and codes a body?

4.4 Aesthetics of Self-Shaping

When we say "body" (*Körper, corpus*), we mean something that is framed and closed, which is also limited by its shell as an object. Everybody is in a particular space. It can even be asserted that spatiality is for the body what time is for Being—an inalienable possibility, reality, and necessity of existence. The distinction between figures and bodies in geometry rests precisely on the assumption that the figure is only an image, and the body becomes a real object in space. All this is still not enough without the existence of a dematerialized substance or "essence" of the body's physicality. Neither the figure as an image nor the body as an object in their mutual relationship is at all conceivable without the relationship of thought and Being. In a metaphysical relationship, a figure appears through the idea or perception of an object.

Thus, spirit always has precedence over matter and ideas over reality. For an object to be created in nature, there must be a possibility for it to exist. Aristotle distinguished between two concepts of shape or form (*eidos* and *morphé*). Since the body appears as an object and as an imagined thing, this double appearance should be understood by the connection/relationship between form and matter. Not a single thing in nature is without a formal-material condition of its existence.

A state of chaos can disrupt the natural order. We then discuss formlessness and meaninglessness, as the middle stage involves the system's disintegration to the point of reaching the zero point of Being. There are three fundamental concepts in Aristotle's metaphysics of the creation of beings from Being, as well as in classical and modern physics. These are form, matter, and energy. Along with form in its two modes, as we have already pointed out, aesthetic (*eidos*) and physiological-psychic (*morphé*), the Being of beings always appears, in its constancy, as a whole or disintegrated, beautiful and sublime or ugly, good and noble or evil and broken. The three fundamental concepts simultaneously determine any possible metaphysics or physics of the body as an object, regardless of whether it is a stone, an animal, or a human. What is enigmatic in its factual necessity? Nothing but the fettering of matter by the form in its singularity. In other words, metaphysics, like physics, begins with the idea of creation and what is created.

One cannot step into the same river twice, said Heraclitus. And this means that repeating the Being in some form outside of its singularity of movement seems impossible. Necessity is, therefore, a kind of absurdity and a wall for thinking that tries to cross the boundaries set by Being itself. What is true for the facticity of movement in space must also be true for the way of appearing in form, which, like the idea of nature, is predetermined and unchanging. The paradoxes of classical metaphysics regarding the body are precise in that it is always determined by something else and signifies something else. Its formal-material structure can change only when the third member of the conceptual order—energy—reaches the threshold of equalization of form and matter.

How is that even possible? Cybernetics answered its second and third orders by setting information storage as a constant event transformation (Hörl, 2008). This not only disestablishes the idea of the permanence of Being in its changes, but also the final process of overcoming metaphysics with the emergence of disembodiment and, at the same time, the technogenetic construction of a new body through "embedding". The answer in cybernetics presupposes a prior clarification of the distinction between "embodiment" and "embedding". What does it embody, and what does it incorporate? The answer seems to be that consciousness enters the body, as cognitive psychology explains. With the development of the brain and the growth of a child capable of simple and increasingly complex thought processes, thinking appears to be a constitutive factor of "humanity". At the same time, "thought" cannot be reduced to mere logical calculation; it also belongs to the broader spectrum of spirituality or emotionality. In contemporary neuro-cognitivism, an interdisciplinary science, embodiment primarily refers to the body's capacity to move and perceive itself as a body.

This applies even under the condition of "implantation" of other people's organs and implants. Mental abilities are not separated from physical ones. In contemporary discussions of the problem of the embodiment of consciousness, the actions of "mental representations and processes" directed at the body are considered: sensations and motor senses (somatic and enactive) (Prinz in Fingerhut, Hufendiek and Wild, 2013, p. 466). We should ask whether other forms of criticism and cognition are embodied. The reason is that the term "embodiment" carries the unfortunate baggage of apriorism and transcendentalism in the mind (*cogito, Vernunft*). It follows that the term is unlikely to be used further because it causes controversy. Already from the fact that the brain is never fully considered a bodily "organ" like the hand, doubts arise about the ontological status of the incorporeal and the corporeal.

The artificial body represents the result of *techno-genetic* construction. However, we encounter the problem of cybernetic determination here. If, namely, the logic of the technosphere stems from the fact that artificial intelligence creates artificial life, then the artificial body appears as a pure mediality of events that can be produced and controlled. Everything that arises from information or digital code must be disembodied before it can be "embedded" in another body. What is this significant change in the determination of the physicality of the body? First, in the process of dematerialization and disembodiment, the body is reduced to a series of functional organs. Formally speaking, the system consists of "phantom limbs"that can be replicated ad infinitum only because their "essence" is in technical reproduction. For the first time, the concept of singularity no longer refers to the unrepeatability of the case of what is alive and irreplaceable.

By contrast, thanks to artificial intelligence, the emergence of new life requires a fluid, mobile body that can function in non-natural environments. It is not only the body that is the object. These are all imaginable constructions of artificial life because their *spatializing* extends to the post-industrial environment. However, what is most important here is the reversal in the "essence" of the concept of an object. The technosphere comprises a network of autonomous objects that think and act according to the logic of artificial intelligence. There are three modes in which they appear: *robot, cyborg, and android*. Moreover, thanks to the change in the "ontological status" of the concept of the object, it seems possible to conclude that a complete reversal of the entire metaphysical scheme of history is taking place here.

"It" as a creature/thing becomes the subject of its own "fate" without transcendental illusions of eternity and immutability. In the process of de-substantialization, the object rises to the level of a self-sufficient and autonomous network of information and, at the same time, "experiences"/"revives" by creatively imitating the irreducible "human nature", with a tendency to transcend it. Gilles Deleuze was right when he wrote in his writings on Foucault that previous forms of existence, such as God and man, are on the way to disappearing into the Overman (Deleuze, 1986/2004, p. 131).

However, this Overman no longer bears a trace of God's face, nor does he feel the sufferings of human historical consciousness in his pursuit of reaching the Omega point of meaning. His "perfection" becomes pure indifference towards the Other. Except, of course, in the execution of program commands as a contingent "essence" of the technical world. It is no longer about the act of "objectifying" the subject, but about the process of "subjectivizing" the object. The dream of machines in the Renaissance era was not just an echo of the hermetic and esoteric understanding of the human body as a microcosm of the universe. Leonardo's machines, as mechanical prostheses of the human body, and Faust Vrančić's parachutes marked the beginning of an intense search for the secret of transitioning into a state of a flying object—the connection between the living and the non-living—through rising above the Earth and travelling on the dark side of the Moon.

Technosphere objects are impossible without visualization or complex digital images. This once again shows the close connection/relationship between the body and the image. The only difference from the analogue image is that the digital one establishes the conditions for what is not present before the world is virtualized. The aesthetic object does not have the status of readymades. On the contrary, its advantage is that it is infinitely replaceable in its plastic singularity, just like the "artificial brain". The image is not a simulacrum of some "natural" source of the sanctity of Being. As a technical body, it is pure information that can, or may not, be transformed into a condition of self-creation of a real object. If, on the other hand, the new understanding of the concept of object shifts the classical metaphysical problem of the mind-body relationship to the primacy of consciousness versus the extension of matter. It is necessary to establish new relationships between the body and the object. Merleau-Ponty, as we have seen, derived the phenomenology of the body on the assumption that the body is an object in space and that human existence is conducted bodily.

Thus, the opposition of mind and body became suspended. However, there is something doubtful in the new monism. For Husserl, the solution was in the intersubjectivity of intentional consciousness. Thus, the *cogito* must necessarily have the property of a noetic act of event creation. Because only man thinks using language. Merleau-Ponty went a step further towards the spatiality of the body as an existential object. It is such an object that is not a thing in the sense of objectivity, but it is not a pure function of a self-positing subject either. However, the problem remained unsolved. In contemporary neuro-cognitivism, philosophy, and science, certainty is sought without unnecessary wandering through the labyrinth.

When here, it is no longer about the "embodiment of consciousness" because the "phantom hand" is already a remnant of the onto-pathology of nature as non-perfection; what else remains but a reversal in the concept of cogito or subject? In Husserl's *Cartesian Meditations (Cartesianische Meditationen)* from 1929, cogito is conceived differently from Descartes's, and the subject becomes more radical than in Kant (Husserl, 2012). The mind and the subject are not, however, the same. What binds them together becomes transcendence. The subject mentally constructs the world through spatiotemporal perception, thereby understanding the essence of their own reality. Everything that can therefore be thought of within the limits of the transcendental subject is determined by the categories of causes of some phenomenon and the purpose it has in the sense of Being. Without causality and purposefulness, Being seems to be meaningless. However, it's not quite like that. Husserl's project of the phenomenological reduction of the essence of the world, in contrast to Kant, is, on the other hand, based on an attempt to break through the enchanted border between mind and nature.

We have seen that he therefore had to leave the body in an environment of intentional consciousness, serving as a mediating force between the demands of the mind and the subject's autonomy. Phenomenology schedules in its program the ripening of pure "beings" exactly where its highest peaks are—in the act of eidetic reduction. It cannot think of the abyss or the groundlessness of that uncanny process of the emergence of the modern *cogito* as a transcendental subject. And it cannot because it starts from the self-evident "fact" that every consciousness (*noesis*) always also appears as the consciousness of something (*noema*).

But what if that "something" (Being?) is the same as Nothing, that is, what can no longer be thought metaphysically, as Heidegger established in the way of his thinking? Should we perhaps abandon this distinction, this firm boundary between "subject" and "object", by simply reversing the situation itself? Therefore, in the concept of embeddedness, the possibility arises that the very process of cogitation or thought criticism becomes an act of objectification of consciousness. At the end of *Cybernetic Anthropology: A History of Virtuality (Kybernetische Anthropologie: Eine Geschichte der Virtualität),* Stefan Rieger introduces the relationship between virtuality and transcendence into the discussion. Of course, we can assume that virtuality in the digital ontology environment cannot be a new *a priori* of technically constructed consciousness. Instead, "virtual transcendence" is at work. And it, on the other hand, arises from the singularity and contingency of events (Rieger, 2003, pp. 422–434).

Isn't the problem precisely in thought's attempt to solve the mystery of the "embodiment of consciousness" using old metaphysical schemes in a new way? As shown by various studies in neuro-cognitivism, consciousness is not located outside the body as an extracorporeal substance that, by the will of God or by an act of spontaneity on the part of the subject, sets in motion the complex mechanisms of human physiological and psychic processes. However, it is also not "in" the body as a mere object that can be disposed of like a pile of flesh and nerves. Thinking as the highest form of conscious activity is a self-reflexive act of knowing the world. The world cannot be located somewhere objectively outside of consciousness. For Kant, time and space were the result of the subject's construction, not an eternal and objective category.

The mind, then, constructs natural laws that do not exist objectively outside of our consciousness. It would be wrong to say that the world is only what is written in the title of Schopenhauer's book, *The World as Will and Representation.* The first energy principle concerns Being in the mobilization of the Earth as a planet, whereas the second appears as a cognitive moment of the subject's rule. To present the world in the medium of thought means to have it as a pure construction of unconditional better. However, the return of realism to ontology at the beginning of the twenty-first century also appears in opposition to a new type of transcendental or radical constructivism, as advocated by supporters of "second-order cybernetics"theories, as seen in the works of Heinz von Foerster and Ernst von Glasersfeld (von Foerster, 1985, von Glasersfeld, 1995). Virtualization means that technical "operational thinking" that overcomes the oppositions of primary and secondary, original and copy, a priori and a posteriori.

What happens to the subject and the object? Nothing but the ontological deployment of their essence. Another important addition: the virtual enables reality to appear at all on the horizon of space and time. Unlike transcendence, which is primarily related to the concept of primordial time, here there is nothing temporal in the meaning of presence as "now". Everything is "here-now". Everything happens simultaneously. Virtuality precedes the actualization of certain states of events that consciousness, in the technical medium of "second and third-order cybernetics", simultaneously produces, visualizes, and "thinks". As we can see, production precedes sight and cognition. The practical character of today's techno-science *extends beyond* this historical development scheme. It is only about the fact that production is not an unconscious act of some complex body in the four causes (*causa formalis, causa materialis, causa efficiens, and causa finalis*).

Poiesis, visio, and computatio are found in a new set of categories and concepts. Instead of the transcendental structure of a subject who thinks by imitating God or by creatively constructing the nature of things, a virtualization of the object's event is at work. In the form of a "thinking machine" (computer), it visualizes the very process of creating artificial life (*A-life*). Objects that think in a way different from human thinking, which Heidegger separated into thinking (*Denken*) and telling (*Dichten*), enter the "third order of cybernetics". They calculate, plan, and construct events as states of affairs in terms of their potential and usefulness. Pragmatics of knowledge replaces the meaning of Being (hermeneutics, phenomenology, and psychoanalysis). When we are no longer able to think of Being and time directly and indirectly, authentically and vulgarly, when we can no longer hope for any "second beginning" (*anderen Anfang*) of thought by a reversal in the "essence" of the metaphysical history of the West, what remains? The answer seems clear: the techno-genesis of events from which the technosphere is created and transformed into a system of thinking, autonomous, and self-moving objects.

4.5 Hybridity, Fractalization, Curvature

There is no longer a gap between *ontogenesis* and *techno-genesis*. This could still be the case at the dawn of the new century, when the first research into the possibility of machines becoming human substitutes, and a man flying from Earth into the endless expanses of space, began. The problem with determining the meaning of an object in the age of the technosphere, in principle, stems only from the fact that it designates a radical construction of artificial life. Then the differences between "subject" and "object" are irrelevant for "operational thinking". Moreover, many terms used to describe the contemporary situation are highly ineffective, if not completely unusable. After Norbert Wiener established the fundamental principles of cybernetics by introducing the concept of information, everything changed, including the notion of objectivity. In the "third order of cybernetics", there is no possibility that the human consciousness would be the one that disembodied itself and thereby acquire the status of a transcendental subject. There are virtual-real objects in the world of technosphere rule, such as devices and apparatus, for which the principle of *autopoiesis* applies. They create themselves in a manner like nature and living systems (biological evolution). Thus, they enter the post-biological stage of "life", for which it is necessary to ensure optimal functioning conditions through the increasing construction of new digital platforms and the increasing incorporation of artificial organs into the assemblies that now shape what we still call the body.

Embedding becomes a means of extending the body as an object, thereby increasing its shelf life—or, in the case of a human, its lifespan, thanks to technoscientific research's potential to alter "nature" as such. Without this, it seems impossible to continue with the utopia of bodily immortality and the dystopia of the end of the human body. Neither the human body nor the animal body is a biologically extinct form of existence or "Being-in-the-world". On the contrary, we will increasingly witness the kingdom of hybrid creations in all areas of human activity and the surrounding world. Being in the form and way of appearing in the manner of a hybrid condition does not mean having a permanent "nature". Instead, everything is subject to technical transformations in response to the demands of aesthetic self-shaping. Desire, therefore, does not come from some sublime "black box" of a metaphysically constructed machine that everyone follows because it is universal, far from it.

The essence of desire lies in its irreducibility to anything in common. Hence, contingency or chance prevails over the necessity (actuality) of Being. The object, in its autonomous action, can adapt to environmental changes, much like a chameleon, or aesthetically construct its temporary environment, as in the experimental process of "terraforming Mars" in the future human conquest of the red planet. Hybrid life also requires hybrid materials, all synthetic, that combine the organism's biological conditions with the cybernetic system for necessary transformations.

To be hybrid means to have an ambivalent experience of the two-dimensionality of a Being that is neither, but honestly that which belongs to the realm of the "Big Third". The order of things in the age of the technosphere derives from the central concept of "third-order" cybernetics. It is an *emergence*. The entire history of complex systems is covered by the emergence of the "new". Why do we write this word in quotation marks? The reason lies in the fact that something new arises from the persistence of Being in change. A new sunrise, a new age created on the ruins of the previous epoch, a new man who knows that it is not only about the aesthetic appearance, but about the spiritual change of his existence (*metanoia*). In contrast, in the world of autonomous objects, "new" arises from the dynamic procedures and protocols by which cognitive machines self-produce their bodies and environments. This is why it can be said that the ontologically new and the cybernetic "new" primarily distinguish the relationship to the historical mission of Being from the relationship to the event as a technical process ("uploading" in transhumanism). In the philosophical sense of the word, it is the relationship between Heidegger's and Deleuze's thinking. Being presupposes the meaning and order of being in the causal-purposive chain of events.

The event, on the other hand, signifies the emergence of structures and processes from the logic of *techno-genesis*, which encompasses the connection or relationship between the living and the non-living. That is why we discuss the rules of creative chaos or the order of nonlinearity. *Emergence* refers to "chance", as existence refers to "necessity" as its natural obstacle in designing the world. The arrangement of hybrid objects breaks down the boundaries between worlds. Thus, the body in infinite becoming (*Werden, devenir*) finds itself in constant transformations of events, and the initiator of this process becomes the cybernetic information system. In the pursuit of reaching "infinite speed" (*vitesse infinie*) in the universe beyond space and time, a new history of post-biological humanity is unfolding (Deleuze & Guattari, 1991/2005, p. 118).

Why are autonomous objects in aesthetic self-shaping condemned to hybridity in all their manifestations, and not only in appearance (*eidos*) and form (*morphé*)? The explanation for mixing substances to create a "new" part of the alchemical process in the search for pure gold seems illogical. Hybridity is the only possibility of Being that which unites events and becoming with a difference. In other words, what medieval theology calls the *tertium datur* in opposition to the rules of "common sense"—such as the existence of unicorns, albino deer, black sheep, or fractally formed cauliflower—refers to something truly ontologically decisive. Instead of the rule that the exception confirms the rule, the leading generative principle is now that the exception determines the rule, that is, that the singularity of what is created from the mixing of two different substances becomes a new way of self-shaping the world. Hybrid objects are media-determined encounters between worlds. The reason lies in the fact that, in the techno-aesthetic mode of Being, they conceal the secret of "new" creation. Hybridity is not, therefore, some external feature of the modern world in planetary-global movement. At its Centre lies the internal structure of the new metastable order. Uncanny (*Unheimlichkeit*) becomes the rule in the construction of the object, much as communication between beings and civilizations in *Star Trek: The Next Generation* (1987–1994) is based on posthumanism, transhumanism, new cosmology, technoscience, robotics, and nanotechnology. The meaning of the word *"hybrid" (ὕβρις)* from its original Greek usage to modern times, particularly in the context of the Technosphere, has undergone significant changes. While it originally had a negative connotation, it now carries a positive meaning.

What seems valid in the biological sense is carried forward in new technologies for the construction of objects of wide application in the daily life of contemporary Human. In doing so, the entire network of newly created technical products is designed according to the aesthetic criteria of surface polishing and fractalization of shapes. Everything becomes curved. Everything is visualized from multiple perspectives. Hybridity reigns inexorably in our lives. Nothing is more self-evident from one dimension and one source. Just as energy is drawn from two or more sources, a form cannot be reduced to uniformity. The metaphorical nature of the body in a hybrid state facilitates its replacement more quickly and easily. However, replaceability and substitutability should not be confused. In the first case, the issue concerns similarity to the original.

In this way, the substitute cannot be a copy, because it inherits certain features of the original. However, his role becomes purely operational. If the replacement toner in the printer works as well as the original, then it is a pragmatic notion of Being. Everything that contributes to the ultimate purpose is good and has its function. In contrast, Derrida calls a substitute a *supplement* in *Of Grammatology*. The meaning of the substitute is as an addition in that, paradoxically, it precedes the Being as a single and singular event. In the world of technical civilization, life is lived as a pragmatic pursuit of goals and objectives. That's why bodies are replaceable like any other object. Their "addition" to what happens in nature denotes the path towards the *posthuman condition*. After all, artificial intelligence precedes the human mind just as artificial life precedes what still seems to us the only worthy of real life.

4.6 Conclusion

Finally, we summarize our arguments to date regarding the transition from the body's ontology to the cybernetic system of information and the appropriate logic of the technosphere. First, nature and the Earth were the foundations and sources (*arché*) for understanding the body as rooted in some spiritual ground. Throughout the history of metaphysics and its transformation in Heidegger, the body has not appeared as an explicit problem or question for thinking; it remains hidden today within neuro-cognitivism under the term "embodiment of consciousness". It could not be thematized separately in its principled autonomy because it had the status and character of a mere object, with the associated features of matter and form (*eidos and morphé*).

From the horizon of intersubjectivity in Husserl's phenomenology, the path to the existential turn and to the openness of the body as an event in Merleau-Ponty's thinking was both the first and the last step in establishing metaphysics through its concepts. Curvature, fractalization, and substitutability are clear evidence that the body, as a living machine, appears in a fundamentally different way than the constant transformation of Being described by traditional metaphysics for centuries. We need not begin by thinking of the body ontologically, as Jean-Luc Nancy asserted in *Corpus*. However, our task is less apocalyptic and messianic than the announcement of the end of history and metaphysics, on which Western thinking about Being has abandoned itself. Instead, it is necessary to begin with the creation of the post-biological body and its subsequent permanent transformations. Why is there such a need for a "new" and changing Being? Simply because there is no longer any illusion that nature and the Earth are the last words of human existence. As interplanetary nomads wander through space, which is yet to come in full scale for the human species, we encounter the monstrous "new nature" of the technosphere, which rests on the logic of the trinity of categories: *calculation, planning*, and *construction*. Hence, the techno-poietic activity of "artificial intelligence"also requires the aesthetic design of "artificial life".

The body becomes a fluid and metamorphic object. But it is no longer an object in the function of some transcendental subject that a priori decides on its movement, form, and material extension in space. As in digital architecture, so too in design for the age of technosphere, one works on designing objects that can independently and arbitrarily, of course, still under the watchful eye of man as a program supervisor without a direct physical presence in real space and time, cross the boundaries set by the actual organization of reality. Getting out of the shelter of nature and the Earth requires the body as a digital object to be miraculously "ascended" to heaven. Considering that the body becomes a task for which we still lack a suitable language. This technique, with which we assemble concepts for hybrid circuits, is full of neologisms and "language games"(*know-how*). But who cares about that if the only truth of language in the age of the *technosphere* is that no one speaks it anymore, except for thinkers and artists lost in the abyss of some phantom primordial Earth, abandoned long ago by both humans and machines?

And maybe forever.

Note The machine is a stranger, and that stranger who precisely creates the human, makes it conscious, materializes, serves it, but always remains outside the horizon of the human. The true cause of alienation in the modern world consists in this ignorance of the machine, which is not an alienation caused by the machine, but by ignorance of its nature and its essence, the absence of the world of meaning and its non-existence in the table of values and in the understandings that have a part in culture. (Simondon,1958/1969/1989, pp. 9–10)

References

Blumenberg, H. (2015). Lebenswelt und Technisierung unter Aspekten der Phänomenologie. In *Schriften zur Technik*. Suhrkamp.

Deleuze, G. (1969). *Logique du sens*. Minuit.

Deleuze, G. (1986/2004). *Foucault*. Minuit.

Deleuze, G., & Guattari, F. (1991/2005). *Qu'est-ce que la philosophie?* Minuit.

Derrida, J. (1967). *De la grammatologie*. Minuit.

Derrida, J. (2010). *La vérité en peinture*. Flammarion.

Freud, S. (1940). Gesammelte Werke. In *Jenseits des Lustprinzips. Massenpsychologie und Ich-Analyse. Das Ich und das Es* (Vol. 13, 11th ed.). S. Fischer.

Heidegger, M. (1954). *Was heißt Denken?* M. Niemeyer.

Heidegger, M. (1977). *Sein und Zeit. GA* (Vol. 2, 14th ed.). V. Klostermann.

Heidegger, M. (2003). Der Ursprung des Kunstwerkes. In *Holzwege (1935–1946) GA* (Vol. 5). V. Klostermann.

Heraclitus. (2013). *The fragments of Heraclitus*. Digireads.com.

Hörl, E. (Ed.). (2008). *Die Transformation des Humanen: Beiträge zur Kulturgeschichte der Kybernetik*. Frankfurt.

Husserl, Edmund (2006) "Grundlegende Untersuchungen zum phenomenologischen Ursprung der Räumlichkeit der Natur" in: Edmund Husserl (1934) Kopernikanische Umwendung der Kopernikanischen Umwendung; in: Jörg Dünne/Stephan Günzel; *Raumtheorie. Grundlagentexte aus Philosophie und Kulturwissenschaften*. Frankfurt/M:. pp. 153–165. https://sdvigpress.org/dox/108091/108092.pdf

Husserl, E. (2012). *Cartesianische Meditationen*. F. Meiner.

Kapp, E. (1877). *Grundlinien einer Philosophie der Technik. Zur Entstehungsgeschichte der Kultur aus neuen Gesichtspunkten*. Verlag Georg Westermann.

Lévinas, E. (1990). *Totalité et infini: Essai sur l'extériorité*. Le Livre de Poche.

Lyotard, J.-F. (1979). *La condition postmoderne*. Minuit.

Lyotard, J.-F. (1988). *L'inhumain: Causeries sur le temps*. Minuit.

Merleau-Ponty, M. (1945). *Phénoménologie de la perception*. Gallimard.

Merleau-Ponty, M. (1958). *Phenomenology of perception*. Routledge. Translated by Colin Smith.

Merleau-Ponty, M. (1968). *The visible and the invisible*. Northwestern University Press. Translated by Alfonso Lingis.

Merleau-Ponty, M. (2007). Eye and Mind. In T. Toadvine & L. Lawlor (Eds.), *The Merleau-Ponty reader*. Northwestern University Press.

More, Max (1993) "The Extropian principles", http://www.aleph.se/Trans/Cultural/Philosophy/princip.html

Nancy, J.-L. (1997). *The sense of the world*. University of Minnesota Press. Translated by Jeffrey S. Librett.

Nancy, J.-L. (2008a). *Corpus*. Fordham University Press.

Nancy, J.-L. (2008b). The intruder. In *Corpus* (pp. 161–170). Fordham University Press. Translated by Richard A. Rand.

Nietzsche, F. W. (1988). *Also sprach Zarathustra* (Vol. I-IV). W. de Gruyter.

Nietzsche, F. W. (2000). *Die Geburt der Tragödie aus dem Geiste der Musik*. Insel.

Paić, Ž. (2011). *The posthuman condition: The end of man and the odds of other history*. Litteris.

Paić, Ž. (2019). *White holes and the visualization of the body*. Palgrave Macmillan.

Prinz, J. (2013). Ist das Bewusstsein verkörpert? In J. Fingerhut, R. Hufendiek, & M. Wild (Eds.), *Philosophie der Verkörperung: Grundlagentexte zu einer aktuellen Debatte*. Suhrkamp.

Rieger, S. (2003). *Kibernetische anthropologie: Eine geschichte der virtualität*. Suhrkamp.

Sartre, J.-P. (1943). *L'Être et le Néant: Essai d'ontologie phénoménologique*. Gallimard.

Simondon, Gilbert (1958/1969/1989) *Du mode d'existence des objets techniques.*: Aubier.

Slaby, J. (2014). Perspectives of critical philosophy of neuroscience. *Tvrđa, 1–2*, 211–221. Translated from German by Boris Perić.

Sturma, D. (Ed.). (2013). *Philosophie und neurowissenschaften*. Suhrkamp.

Virilio, P. (1991). *The aesthetics of disappearance*. Semiotext(e).

von Foerster, H. (1985). *Sicht und einsicht: Versuche zu einer operativen Erkenntnistheorie*. Springer.

von Glasserfeld, E. (1995). *Radical constructivism: A way of knowing and learning*. The Falmer Press.

CHAPTER 5

How Else Can We Think About Art? Figurality, Aisthesis, and Visualization

5.1 Introduction

Schlegel's twelfth fragment from 1797 argues that what is usually designated by the concept of philosophy of art lacks one of its two components, either philosophy or art. The matter could still be remedied if there is too much of one and too little of the other. But what if this fragment of Schlegel and its critical aphorism is nothing more than a good joke, or if, to make matters worse for him, it is no longer a relevant thing directed at philosophy and art? Perhaps it would still have been interesting had Hegel and Schelling constructed their philosophical systems, thereby elevating metaphysics to the pinnacle of modern philosophy.

At that time, it was self-evident that philosophy as a fundamental ontology, as Heidegger would say, should speculatively and reflexively address art from on high to give it a foundation as a kind of new discipline within the philosophically understood *metaphysica specialis* Thinking about art in the era of the emergence of transcendental systems and Romanticism was already marked by an apparent controversy—namely, philosophy since Kant had been impregnated with rationalism and the language of natural and technical sciences. The Age of Enlightenment was necessarily guided by the idea that any mysticism and mysteriousness of nature, as well as that of human superstition and religious dogmatism, take precedence over the scientific method and research into being

Ž. Paić, *The Technoscientific Turn of Philosophy*,
https://doi.org/10.1007/978-3-032-21081-4_5

and the essence of man. Kant, as is known, did not establish any philosophy of art in his *Critique of Judgment*, nor did he develop philosophical aesthetics, because Baumgarten had already done so. The entire problem with the emergence of aesthetics as a reflection on the beautiful and the sublime aspects of the natural and human worlds of art was best summarized by Hans-Georg Gadamer.

> Aesthetics as a philosophical discipline emerged only in the 18th century, i.e., in the era of rationalism, obviously provoked by modern rationalism itself, which rises based on constructive natural sciences, as they developed in the 17th century and as they determine the face of our world to this day, so that they are transformed into technology at an increasingly dizzying pace. (Gadamer, 2003: p. 27)

However, aesthetics, as a philosophical discipline, and the aesthetic, as a way of thinking about sensibility through which the world appears in the horizon of human thought, have something else that is extremely important. The Italian semiotician Umberto Eco primarily aims to liberate us from ignorance about the Middle Ages and, thus, from the mistaken belief that scholasticism was inferior in these matters because transcendental beauty was often relegated to a secondary position. God is first and foremost One, sound and valid, a being of pure perfection, and only finally does He appear to man through the splendour and veil of beauty. Therefore, starting with the "founder" of philosophical aesthetics, Baumgarten, Eco lists all its meanings: the science of sensory cognition, the theory of the liberal arts, the art of beautiful thinking, the *art of what is analogous in reason, including* industrial design, life experience (happening), experimentation in language (Joyce), or some combinatorial manipulation of elements (art produced by computers (Eco, 1988: pp. 1–20).

Isn't it notable that, back in 1970, when Eco published this book in Italian, he was already sure that computer manipulation of elements would become an art. Today, many respected aestheticians will not admit it because of what powerful *AI does* with its algorithms. We see that the difference always places the aesthetic in the realm of nature, even though crime represents an act of "spiritual nature". At the same time, we perceive it artistically from the connection of spirit, society, life, language, and technique. However, what is missing in Eco's enumeration of Baumgarten's terminology lies beyond the traditionally understood domain of aesthetics. In the middle is the insight that Heidegger and Gadamer mention in their discussions of criticism of modern metaphysics, especially aesthetics, claiming that the eighteenth century, when aesthetics arose, was the age of the birth of rationalism and contemporary technology. The machine that determines this *mathesis universalis* denotes the beginning of the mechanical era, with which a complete acceleration of life occurred.

Art in its profane form became nothing more than an aesthetic shape of the ethical virtues of modernity that Schiller celebrated in his *Letters on Aesthetic Education* (Schiller, 2019). Everything that aesthetics meant at its origin no longer means today. However, what remains lies beyond any epistemological or theoretical relativism, so close to the postmodern critique of metaphysical universalism. Beauty cannot be immortalized because it is not primarily a matter of computer art, but rather something that transcends the boundaries of both the aesthetic and the artistic, and appears in entities not as an illusion of truth and goodness, but as a synthesis of technical and human activity, and a synthesis of nature and spirit. Therefore, that's all the epiphany and transcendental, construction and existential in a completely different notion, unlike the traditional metaphysical one.

5.2 Figurality

The philosophy of art that has emerged since then has left no doubt that what is proper to art—its ontological groundlessness and cognitive-reflexive intuitiveness—cannot be thought irreducibly without a radical critique of philosophy, specifically rationalism, and philosophy as the foundation of art. German Romanticism, with Novalis and Schlegel, found an alternative in a return to the pre-reflexive, to that which lies beyond the *logos* as the foundation of Being in general. Hence, it is no coincidence that the most significant breakthroughs within and outside the philosophy of art were made simultaneously in the writings of Nietzsche and Kierkegaard, as critiques of Hegel's absolute metaphysics and Romanticism.

Many will nevertheless add that the most significant philosophy of art of the nineteenth century is that of Schelling, from which even today we can arrive at the possibility of finding a condition of possibility for the system of logical foundation of the science of art that is neither philosophical nor theological, but arises from the abyss of the world as that irreducible artistic action outside the logic of causality and purposiveness (Welsch, 1993). However, it seems that Nietzsche and Kierkegaard are closer to attempts to renew philosophical thought in its essence, beginning with the nature of art rather than with any scientific-technical rationality, on which, as Gadamer has shown, philosophical aesthetics is also based. In this regard, Nietzsche required the philosopher to think and live artistically, whereas, in his early phase, Kierkegaard established art as an existential necessity for guiding human freedom, thereby marking it as the modern destiny of man.

I addressed this problem in articulating the concept of the technosphere and concluded that no new philosophy of art can provide a salutary insight into, for example, the achievements of trans-gen art or post-conceptualism, let alone possess an adequate conceptual-categorical apparatus for understanding the problem of the body as a techno-genetic construction (Paić, 2021, 2022). Why? This is because the philosophy of art cannot conceptualize what the concept of event already signifies in Heidegger, let alone in Whitehead and Deleuze. For Nietzsche, art designates already an event of the will to power as an aesthetic becoming (*Werden*), and not a permanent and unchanging Being to which beauty and sublimity, as in Kant and even Hegel, are bestowed by God as a regulative concept of the purposiveness of nature or as a theodicy of the world spirit.

Therefore, art after Heidegger's requiem for ontology, *Being and Time,* before Deleuze's *Difference and Repetition*, must become the existential-aesthetic project of the synthesis of life and that which transcends life. This is evident today in artificial intelligence, as seen in ChatGPT, an Open-Source product developed by OpenAI. What should this be other than a fascinating artistic event that synthesizes mind and will, aesthetic appearance and self-reflection, the autopoietic and the autogenetic in all forms of possible existence within this world? What would any hermeneutics and phenomenology, in a Gadamerian-Husserlian tone, have to say about this paradigmatic case of proof that the relationship between philosophy and art in the age of the technosphere means the end of philosophy and the end of art within the limits of Western metaphysics?

Heidegger was the first to clearly warn that the time of the reign of the enframing (*Gestell)* as the essence of technology is coming and that the only salvation of what belongs to the nature of philosophy and the essence of art lies in the event (*Ereignis*) as the place of the openness of the meaning of Being which cannot be Parmenidean the same and unchangeable, but appears in the event and postponement of its possible, but not necessary, happening. Nothing is any longer fundamental or essential as it was for the philosophy of art and the aesthetics of the Kantian-Hegelian circle of thought. Still, nothing remains only a possibility. However, insofar as this possibility is understood as potentiality, it cannot simultaneously serve as the condition for the emergence of the "new" as the fundamental concept of avant-garde and contemporary art. Nothing "new" is any longer accurate or necessary, and perhaps it is only possible if a complete reversal occurs in the very essence of the post-metaphysical notion of philosophy and art.

What remains is "usurped" by different, so-called new aesthetics of performativity, installation, atmosphere, image, digitality, and corporeality, as they have shifted from Deleuzian thinking in terms of eschatology to analysing the brain as the creator of thought events through fractal virtualization. From this, we are no longer talking about philosophies of art or philosophical aesthetics. Still, *aisthesis* comes into view only in the plural world of the aesthetic Being, which ranges from Max Bense and cybernetics to Dieter Mersch, in the irreducible openness of human creativity versus the technological reduction of *AI creativity* (Bense, 1965; Mersch, 2015; Paić, 2022). It concerns the relationship between figure and concept in the practical use of artificial intelligence tools in creative activities, such as writing in the humanities, literary works, and the composition and performance of musical forms. Namely, we see that it is becoming more of a question day by day whether creativity, as the essence of human thought activity, as Deleuze also does in his philosophical texts, especially in the late stages, is threatened by the technosphere, *which* has the possibility of autopoietic experimenting with language, image, and sound in creating the "new" (Deleuze & Guattari, 1991).

In the book *Discours, figure*, Jean-François Lyotard performs figurality in thought in an exceptionally striking way. These figures analyse Klee's painting and his instructions on how modern drawing should be shaped by reducing content to lines, colours, and figures in space (Lyotard, 1971). Conceptuality in thinking, by contrast, in the pictorial organization of thought, becomes a key site in Deleuze's ontology of becoming (*devenir*). Figures and concepts within the framework of poststructuralist philosophy played a crucial role in challenging the classical metaphysical ideas of transcendence, particularly regarding the primacy of language. Furthermore, it is not at all accidental that both terms are derived from the effort to liberate thinking in the face of technologies of reproduction and their powerful diapositives of repetition and simulacrum from the logic that characterized Wittgenstein's *Tractatus logico-philosophicus* and which he soon abandoned.

It is a transcendental structure of thought that a priori determines the limits of language. However, what Lyotard does with the concept of figure, but also Deleuze in his book on the painter Francis Bacon, shows that we must bear in mind that this non-rational process of thought in painting and film, which provokes an even more pictorial rupture of affectivity because it addresses the masses, is not predetermined by anything. After all, Deleuze insists that Bacon's authentic act of painting was done without sketches (Deleuze, 2005). In cinema art, the actor changes the script on the spot because the very event of creating something new has a contingent "necessity" in itself. Therefore, figural thought appears as creatively determined from the indeterminacy of intuition. Consequently, a figure always represents free formation in space-time, but within the limits of what gives the form its formativeness—this attempt to follow a conceptuality in thought in a correlation between Platonism and contextualism.

The form in which the figure appears testifies to the correlation of the idea and its appearance. Hence, painting should be understood as an expression of figural materialization grounded in the creative principles of autopoiesis, because the image, as information, constantly produces other information. The concept is what gives the image meaning within a specific context. That is why artificial intelligence, like *a third hand*, can paint the object of figurality, encompassing the noematic field of the conceptual openness of the art of presentation. How does it do that? By mimetically and representationally imitating and constructing what already exists, plus adding one's artificial creativity, which belongs to experimentation with numbers, based on post-memory or files of information about a painting style. Namely, figurality is spatially determined, and conceptuality stems from the idea of time as *nunc stans*.

This means that thanks to the programmed set of calculated images, as the new media theorist Friedrich Kittler would say for the concept of the digital image, (Kittler, 2005: pp. 186–203) it become possible for a "thinking machine" to produce a new reality as a creative simulacrum in the shortest possible time, just as Lyotard and especially Deleuze conceived in their ontologically innovative reflections on the relationship between image and language of thought. The same applies to linguistic forms in the sense of literary reflections. *AI* will soon "write better" than Houellebecq or Rushdie because it will combine the experiences of Joyce and the new novel, provided that the prose-writing program includes a few key words or syntagmas. Figurality and conceptuality are interconnected, contrary to the misconception that there exist two distinct forms of thought. Heidegger refers to these as *Rechnen* and *Dichten*.

AI is based on the logic of "rational intuition", a combination of concept, figure, and sound, despite its metaphysically fluid meaning and mystification of the meaning of music, is mimetic *par excellence*, so that already on *YouTube* we have a melody sung "more perfectly" than the original and authentic one, for example, some Neapolitan aria from Verdi's operas. The technosphere denotes artificial life and a phantom of organic and inhuman synthesis. That is precisely why it is necessary to see how and in what way humans, as creatively irreducible, can still and under what conditions survive in thought without becoming a mere technical artefact, a *readymade*, an object without the sublime within it?

In the contemporary discussion of the possibilities of aesthetic thinking, the German philosopher Dieter Mersch must undertake an exceptionally stimulating reflection on contemporaneity, focusing on three decisive concepts: *aura*, *event*, and *aisthesis*. The first two are inevitably linked to Benjamin and Heidegger. At the same time, the third denotes a newly created concept that breaks with the tradition of modern aesthetics and seeks to foster openness to meaning regarding the connections among technology, artificial intelligence, and creativity. My research aligns closely with his premises in this regard (Mersch, 2002).

Unlike his attempt, the technosphere concept results from autopoiesis and the artificial intelligence of third-order machines, namely cybernetics. So, I conclude that new AI devices, by the very logic of things, will no longer be copies and simulacrums of the human in terms of technical mimesis and representation, but rather in the creation of new as synthetic in something that goes beyond the history of art to date, not of course in the sense of "better" and "higher quality", but "aesthetically more efficient", hyperplastic and performatively more fascinating than anything seen so far. The problem that Mersch analysed in detail concerns the essence of thinking, the so-called creativity of artificial intelligence, which raises questions about the human mind and its achievements, particularly in the mystery of creativity. Be that as it may, it seems necessary to rethink the limits of the language with which human thinking relates to the technological environment, primarily because the non-human or artificial thinking of a machine, no matter how intelligent, cannot have within itself what Lyotard called a *conceptual hologram* or openness in all directions.

5.3 Aisthesis

Aisthesis denotes an undeniable, truly stimulating philosophical concept. What is natural and technically sensory cannot be a storehouse of pseudo-emotions, in the sense of programmed responses to externally elicited stimuli, such as sadness, joy, suffering, pleasure, anger, and contempt. Human corporeality encompasses sensory hyper-plasticity, just as the human brain does. The fundamental difference lies in the "ontological" difference between vision and program, "embodiment" and "embedding". When something can be programmed, it follows computational logic. When something has a vision, it becomes, in its ultimate intention, creatively open to the possibility of emergence. The program's language becomes *know-how*, and the language of vision is precisely this *aesthetic*, a telling horizon that unites philosophy and art (Paić, 2023a).

However, it seems that the real question should be how long can this creative language of events, from its aesthetic perspective, be an alternative to the language of programs, to that empire of bare pragmatics that moves forward at such speed and destroys everything before it, including the world of natural languages, which it reduces to the question of Leibniz monadological infinity starting from the logic of 0–1? Number is nothing more than a pure form in which Being becomes thought of as the creation of a being from Nothing. Isn't Roger Penrose, therefore, justified in asking whether mathematics represents the work of God's providence or, rather, our constructive invention of the idea of the world and its rules and laws? (Penrose, 2007: pp. 12–17).

Can we take this one step further and consider what it would mean if it were purely contingent? What with numbers and mathematics, it simply emerged as an event, neither from the mind of the immovable mover nor the human imagination, but from pure chance? Namely, all concepts of traditional philosophy of art and aesthetics, as we have already seen in the Gadamer quote, are imbued with the "impurity" of modern scientific and technological constructions. If that is so, it was necessary to understand that the future of the aesthetic cannot be outside the technological one, even if they put themselves on a head made of obsidian. The matter seems so simple that the head must ache not from the uncanny complex problem, but from the most complicated possible simplicity. The concepts of philosophy, art, and aesthetics are no longer applicable. The journey with the technosphere is complete. What will still "happen" as "new" will not be anything new, because newness denotes the repetition of actualization in the temporal structure of the "now", which only extends like the Cartesian *res extensa*.

New might be determined as the obsolescence of this world and its eternal novelty, a concept unknown to the ancient Greeks. Socrates, after all, mocked an Athenian who travelled everywhere and claimed to have met "new" people and places, without knowing or realizing what old people and places were, let alone what new ones were. When the technosphere becomes an artistic event of interactive visual communication, we are in an accelerated era of artificial intelligence, and it is no longer a *work of art*. The synthetic is nothing fixed and unchangeable, nor static and permanent. Rapid change and transformation characterize it; therefore, neither art nor aesthetics can lack reflection and speculation. Because *ChatGPT–OpenAI* tends to think by programming its vision and speak by algorithmically calculating and planning constructed states, thanks to artificial language as the technological device of all existing dead and living languages of human civilization.

5.4 A Picture Without a World

The most significant philosophical treatise of the twentieth century is Heidegger's "The Age of the World Picture" ("Die Zeit der Weltbildes"), published in 1938. The reason is apparent and already stated in the title. Time, as the age of the world picture, has always been present throughout the history of Western metaphysics, marked by its epochality. It is therefore not some immobile eternity of the permanence of the same, but the historical world-making of the world as the rule of that complex of Being and time which is no longer anything open in the sense of the historicity of "nature" and "Human". Instead, we encounter the rule of Being as the established objectivity of the object, which encloses the whole of nature and the Human in the complex scientific-technological thrust of thought. All of this begins with the transformation of the Greek concept of *hypokeimenon* into the Latin idea of *subjectum*.

When the subjectivity of the subject appears on the horizon of modern representation of the world, the original openness of the world, to which meaning was given by thought as telling (*Dichten*) in the sense of the mythopoetic secret of the encounter between gods and Humans, nature and the non-human, is over. The modern world, where the subject's rule over the original appearance of Being, beings, and the essence of Humanity unfolds under the power of thinking characterized by calculability and representation, emerges. Calculating thinking becomes thinking "about" nature in its representability. It does not appear as an original world phenomenon, but as a scientific and technical work of research and transformation of the appearance and essence of Being. According to Heidegger's analyses, this process unfolds so that we no longer have the world before our eyes in the purity of its openness, but rather in the worldliness of the world. The modern world represents a radical and fundamental shift in thinking and Being, which is now constructed from the perspective of the world.

From the beginning of his treatise, Heidegger shows why metaphysics has this unconditional power to determine and shape the essence of Being in this distinguished epoch of the historical conception of Being. Namely, "science belongs to the essential phenomena of the modern age" (Heidegger, 2003: p. 69). However, science cannot be something in and of itself, independent of giving Being a different meaning and place in the world. The rank of this modern-age-specific science lies in the omnipresence of the mechanical technology (*Machinentechnik*), which for Heidegger does not mean "the mere application of modern-age mathematical natural science in practice" (Heidegger, 2003: p. 69). Mechanical technology inherently involves a reproduction logic, transforming nature into a means for other purposes. The problem with the instrumental notion of technology, as it evolved through German philosophy of technology from Hegel to Marx to various currents and orientations of the Frankfurt School and Critical Theory, is that the essence of technology cannot be derived from the changed nature of modern science. Instead, it is organologically assumed that science remains an insight into the logic of Being, which is permanent and unchanging. It denotes the natural-scientific structure of thinking about nature as something existing in the sense of objectivity of objects.

It must be admitted that Heidegger shows how this "mechanical technique" in its instrumentality necessarily represents a visible "essence of modern technology". What does this mean? Is the essence shown or hidden if the technique that stands in the light of mathematical natural science as its flywheel and driver is something other than the mere application of science in practice? Heidegger, therefore, shows how the presentation of Being in the modern era becomes essentially determined by the appearance of "mechanical technology". This applies not only to the practical world of nature but also to that which emanates the concepts of beauty and sublimity, that is, to art. The change Heidegger observes is of far-reaching significance for understanding aesthetics as a philosophical discipline that emerged at the end of the eighteenth century, alongside rationalism and the influence of natural and technical sciences. This applies equally to art itself, which is no longer under the authority of the traditional concept of imitation (*mimesis*) of Being.

Still, the idea of aesthetic appearance emerged in the twentieth century with the historical avant-garde movement. It is about representation (*repraesentatio*). When the image is mimetic, it always occurs in the natural environment, reflecting the openness of Being. When the image appears as a representation of the world, then the world must be scientifically and technologically constructed to represent precisely this new thinking that unites nature and mechanical technology. Such thinking determines a modern blueprint and outline of what is not Being as such and in its entirety but is produced by thought or by scientific and technological construction. For Heidegger, art changes significantly in its difference from Greek art in such a way that "the work of art itself becomes the object of experience, and accordingly art becomes the expression of human life" (Heidegger, 2003: p. 69, Paić, 2021).

After the emergence of science, mechanical technology, the aesthetics of experience and art as expressions of human life, the fourth phenomenon of the modern age that has far-reaching significance for modernity is the emergence of culture through "the realization of the highest values…in the essence of culture lies the provision of goods…and the emergence of cultural politics" (Heidegger, 2003: pp. 75–76).

Is this rank of metaphysical reversal in difference from the Greeks simply the result of the causal-teleological scheme of the development and progress of Western history? In "The Age of the World Picture", Heidegger, by clarifying the derivation of thought as the conception of Being, arrives at something genuinely decisive for the further philosophical analysis of modernity. Modern culture, therefore, cannot be something autonomous and magnificent. Still, a drive made possible by the transformation of the Greek paradigm of knowledge into modern science, which, with the help of mechanical technology, conditions the emergence of the new in the sense of a Being that becomes the work of the Human and as such a matter of the provision of "cultural politics".

Science and mechanical technology are the conditions for the possibility of the emergence of the "new".

In this context, the aesthetics of experience transform into the enjoyment of beautiful objects, which must be preserved from the destruction of time through state cultural policy measures. Both the causal and the purposeful, science-technology and aesthetics-art-culture, are nothing other than the already established order of the modern reduction of the world to a picture of the world, beginning with the shift in the concept of image from *mimesis* to *repraesentatio*. However, the image in question here arises from the "image of thought", so the image as an object that is observed and that observes us is, as Paul Klee would say, what now connects no longer Being and thinking, but thinking and Being in the sense of a pure technological construction. Finally, what does Heidegger include in his analysis as the fifth phenomenon of this metaphysical reversal of the relationship between Being and thinking with the emergence of the modern picture of the world?

Nothing other than the emergence of "un-deification" (*Entgötterung*), but not as "crude atheism" (Heidegger, 2003: p. 76). I will not go further and deeper here into the thoughts that Heidegger probably expresses most radically in contemporary philosophy about the loss of the rank of religiosity and the total theological inability to think through this event as something of crucial importance for the future of faith in general and Human as a religious "subject". I will only say that "un-deification" becomes an act of the final process of the disintegration of Western metaphysics that ends with Nietzsche's statement about the death of God as the rule of nihilism, in which what Heidegger calls "the emerging void that is replaced by the historical and psychological research of myth" necessarily comes about (Heidegger, 2003: p. 76).

The act of "un-deification" is not, therefore, a matter of a human-to-human worldview-political decision to bestow upon the modern age both active-passive nihilism in the form of atheism and religious fundamentalism, both of which are perpetrators and witnesses of the same Janus-faced process of denying the divine and its furious and militant reintroduction into the world of secularization and profaneness. Heidegger, therefore, shows that the concept of the "new" in the sense of its unconditional rule and power cannot be a matter of any mere anthropological reduction of metaphysics, but of something essentially "deeper" than what appears in the "new" phenomena. The "new" cannot be understood without a thinking that rests on the image as a representation of the object's construction in its objectivity, and without an answer to the fundamental question about the essence of modern science. The modern age should be understood, therefore, as the "age of the picture of the world", and this image has already been significantly changed in its essence because it does not mirror the world as such in the sense of imitating its permanence and durability. Instead, the image becomes a matter of representational thinking, with its essence residing in its construction (Paić, 2022).

Let's stop here. It seems needless to repeat that in this discussion as a lecture, Heidegger has come furthest of all the philosophers in the circle of the end of metaphysics from Hegel through Nietzsche to Wittgenstein. Only Gilles Deleuze in his "ontology of becoming"established the image in a completely different way as an "image of thought" and that it was not by chance that he saw from the logic of the cinema event how the progress of information technology in recording technology (image-movement and image-time) and the understanding of what I call the technosphere as the third order of cybernetics or *autopoiesis* are a signpost that points to two fundamental concepts of contemporaneity. They arise from the technoscientific turn as *quantification* and *visualization* (Paić, 2023b).

In Heidegger's reflection on the image, which is rooted in the paradigm of mechanical technology, it becomes obvious that a reversal has occurred within the causal-teleological direction of metaphysics itself. It is no longer science that determines the "what" of technology as modern technology, but the "how" (in medieval ontology, we call this the primacy of *quodditas* over *quidditas*) that is crucial in the logic of research that becomes the "essence" of the entire turn in the very essence of science. We see, therefore, that already in 1938, in the context of the "Age of the World Picture", Heidegger takes the most crucial step towards his fundamental concept that ends the period of ontology in general: the enframing (*Gestell*) as the essence of technology. Of course, this will happen later in the lecture "The Question of Technology" from 1953 (Heidegger, 2009).

But one thing should be undeniable. Moreover, it remains a subject of unfinished reflections. Suppose the question of the essence of modern science is impossible to resolve without insight into the nature of contemporary technology. Is it impossible to derive the question of the image as the essence of the technosphere from Heidegger's thinking, which he so inspiringly brought out in 1938 in "The Age of the World Picture"?

My answer would be yes. The limits of thinking that speaks of "Being", "metaphysics", "representation", "image" are the limits of thinking that overcomes metaphysics as the rule of cybernetics with a fundamental concept in five phenomenal articulations in which the image appears as a condition for the possibility of the emergence of (1) science, (2) mechanical technology, (3) the aesthetics of experience and art, (4) culture as cultural politics, and (5) the un-deification of the world in the void of a substitute for the original myth and religious experience.

These limits are precisely determined by the triad of what does not align with Heidegger's paradigm of mechanical technology, which he adhered to until the end of his life, in the changes in the essence of technology from atomic to nuclear. But his notion of cybernetics never reached the last "twilight zone", namely, the reign of quantification and visualization of the technosphere itself, which transcends the limits of the causal-teleological notion of the world. The technosphere, in contrast to mechanical technology as a modern technology of reproduction of "Being", "beings", and "essence" of Human, cannot be longer an image in the sense of *mimesis* and *repraesentatio*, but should be determined from the triad of calculation-planning-construction of "artificial reality" which, with the help of numbers, creates an image as a visualization of events in its contingency-emergency through the cybernetic fourth of information-feedback-control-communication (Paić, 2023c).

What does this mean other than that it seems impossible to think of the image as the essence of the technosphere philosophically within the framework of thinking of overcoming metaphysics as ontology, because the image cannot be longer constructed from the representative thinking of the subject. Instead, the image denotes the autopoietic transversal logic of the rule of autonomous objects that think by seeing themselves in the same way that the god Dionysus looks at his eye, as Peter Sloterdijk effectively expresses in his study of Nietzsche (Sloterdijk, 1986). The problem is, therefore, that the idea of the "age of the world picture" in its magnificence is still only the most significant reach of an image as a conception of the world in the horizon of the end of metaphysics. Even when the possibility is opened that the new paradigm of science represents the one that Heidegger had in mind at the end of the 1960s in conjunction with atomic and nuclear physics, the image has not yet become an autopoietic "time machine" that creates contingent events, rather than mirroring and representing an already existing reality.

The image can no longer be thought of philosophically, because it transcends the limits of metaphysics as a horizon that presupposes a distance between event and reflection. The problem cannot be in the image itself, but in the thought itself. If thought lies in the extra-metaphysical sense, it signifies the possibility of characterizing work or structure of that fourth without which there is no metaphysics, namely, Being-God-World-Human. It is not a matter of the mere disappearance of the framework or structure of metaphysics, but of the disappearance of metaphysics. This means philosophy is no longer a love of wisdom or absolute knowledge (Heraclitus and Hegel).

What remains of philosophy in contemplating the image as the essence of the technosphere is only the "image of thought" in the language of the conceptual event that creates the "image-ness of the image", which no longer has its object or image in the so-called reality. The image has been constructed since the modern age using mechanical technology. In the twenty-first century, the image should be created as a visualization of an event, rather than a representation or a copy of reality in the sense of objectivity. Of course, the mimetic and representational functions of the image thus lose their relevance for contemporary thinking of the technosphere.

5.5 Visualization

Mechanical engineering has become obsolete. We live in an era of techno-scientific construction of "artificial worlds", shaped by the binary logic of quantification and visualization. And that is why we no longer have pictures of the world. We observe the emergence of new probabilistically created worlds as contingencies in which completely different "concepts" and "categories" rule than those analogous to those with which Heidegger meant the original notion of *Being and Time*. It is no longer an "image" because its "essence" lies in the realization of cybernetics in the technosphere, and its imagery cannot be understood starting from ancient historical-epochal "images" as symbols and metaphors of a world that may have passed forever. The art we call "art" by habit has become the aestheticization and culturalization of "artificial life", and only fluid, changeable rules, such as the simulacrum of life, apply to it. We do not have an image of the world because our so-called age arises from the metastasis of the hyperreality technosphere that neither imitates nor represents anything.

It simply constructs new states and contingent events; hence, we are indifferent to its "pictorial" disappearance, just as we are to the image of the Roman statue of the god Apollo, the first valid deified symbol of vanished beauty in the ruins of a bygone time. What we still call "image" results from the cybernetic production of events as the fourth of information, feedback, control, and communication. Man is not the producer of images, but the agent of the process of their illumination, because the meaning of an image has historically always been to depict the splendour and sublimity of its referential frame. Therefore, the classical age of the image was characterized by the prevalence of myth and religion in the Greek and Christian Middle Ages. With the new Era and the age of the scientific-technical "image of the world", the process of the end of metaphysics begins, and in its place, in the mid-twentieth century, comes the cybernetic system of the technosphere.

Since it is no longer about technique or technology, because the technosphere cannot be a means to an end, nor an end in the sense of representing the world of modernity, the technosphere signifies the transformation of "Being" into "becoming" by the world as calculation, planning, and construction of artificial worlds from the essence of artificial intelligence. What is happening here, beyond the establishment of a complete nihilism, is the emergence of a nihilism in which it is no longer a question of the rule of the Overman in the Nietzschean sense, but instead of the nihilism of autopoietic, autonomous machines that think and shape new virtual worlds from the essence of simulation. The age of "pictures without a world" becomes an era of contingency, entropy, and chaos, in which neither philosophy nor art can serve as a relevant discourse. Therefore, it seems entirely wrong in the horizon of thought to continue to prolong the life of philosophical aesthetics that have already lost credibility with Adorno.

In the real world of artificial intelligence, the aesthetic is nothing other than the technosphere that, like the eye of Dionysus, sees its image in the fractals and transversals of an ever-new reality. However, something else becomes highly intriguing in the issue of belief or loss of faith in the power of images. It is precisely this excess of theological-religious "thinking" in the concept of image that appears from the very beginning in the Greeks, and continues throughout the entire history of Western metaphysics, even in the notion of the technical or digital image in Flusser, who, like the late Horkheimer, modernized his Jewish eschatology and messianism with the idea of a telematic society for which the transparency of the image signifies the possibility of merging with the divine (Flusser, 1998).

In *The Iconoclasm of Contemporary Art Aesthetics and the Iconoclasm of Contemporary Art—The Pictures Without the World* I speak of the end of the image not only as the end *of mimesis* and *representation*, but also of the end of the image in the sense of any sublime remnant of the divine in the image (Paić, 2021). With the concept of immanent transcendence, I sought to demonstrate the impossibility of the image becoming a new language, as such a move would merely be the continuation of logocentrism by other means. The real problem with this notion of the image as a technosphere is the disappearance of the traditionally understood space and time of its form of appearance, rather than appearance itself (Purgar, 2019: pp. 106–121).

My theoretical position regarding the *iconic turn* was to find what medieval theology called *tertium datur*. Thus, both language and image are enabled by something synthetically and analytically autonomous, the very "thing" of thought that becomes, from the aesthetic object of Marcel Duchamp to the autopoietic activity of Ken Rinaldo's robot, the same thing as self-producing and self-moving, but so that there is no longer a difference between idea and appearance, transcendence and immanence, but instead of *linguistic and iconic turn* now works on the visualization of the concept as a technosphere. Power no longer has images, but rather what directs the image from the technical dispositive of power to the power of number, or rather the mathematization and technologization of Being and time, to use the language of Heidegger from the *Sein und Zeit* period.

The technosphere designates the rule of the absolute triad of calculation, planning, and construction, and can no longer be understood in traditional metaphysical terms, whether in language or image. Admittedly, the late Wittgenstein sensed this when he introduced the concept of language games (*Sprachspiele)* into philosophical discourse as forms of life (Wittgenstein, 1953). However, here we are discussing pure Platonism in reverse. The idea that the *technosphere* auto-poetically visualizes the world as an image stems from its hyper-plasticity, as expressed by artificial intelligence. Hence, the image in the post-digital world, which presupposes precisely the power of numbers and the mathematicalization of the world as a metaverse, requires the abandonment of philosophy, theology, and art history. Let us go one step further: *Bildwissenschaft* and *Visual Studies* (Sachs-Hombach, 2005).

Why? The answer also presupposes an answer to the question, "Do we still believe in the power of images?" Our so-called faith was not a question of faith in the secular god of information as a condition of the possibility of cybernetics. So, when I titled the book Aesthetics and the Iconoclasm of Contemporary Art with the subtitle "Pictures Without the World", I wanted first and foremost to follow in the footsteps of the late Heidegger and his concept of *Kunstlosigkeit* to reach the other shore, one where there is no longer any difference between art and non-art, but also no difference between living and non-living. That is why the concept of the technosphere denotes an era that transcends the idea of art, from the Greeks to Hegel and beyond. Art, like science, exists in the age of the *technosphere* and is only possible as a research and experiment in the emergence of the new from the spirit of *autopoiesis*.

This spirit, metaphorically speaking, denotes that thinking which itself produces its reality no longer as the difference between idea and phenomenon, *noesis and noema*, the Kantian-Husserlian discourse. Such thinking produces beyond any knowledge of the image in the sense of philosophy or semiotics, as was necessary for the science of images. Science and socio-cultural perspectives on image representation in visual studies encompass diverse viewpoints, including insights, reflections, and observations. In the tradition of Platonic and Plotinian theology or metaphysics, the image has always been a flash of God's light and a reflection in the observer's eye of what is invisible. Thus, the allegory of the cave marks the beginning and end of the aesthetics of appearances. We have nothing to do with that anymore, because synesthetic thinking produces visualized objects as fragments of a contingent and emergent *mappa mundi*, thereby forming a new reality without the need for an external creator.

The end of the so-called faith in the power of images is not the end of philosophy and art in the face of the triumphal march of the technoscientific "image of the world", which will soon accelerate as soon as quantum computers start operating. After all, the fundamental issues of today are not decided by philosophers, theologians, or artists, but only by a triad of astrophysics, cosmology, and biogenetics. This means that thinking speculatively or reflexively, with the image as its subject, must become transversal and experimental in the face of a radical shift in the concepts of power and image in general. Power is not merely political power, but rather that which conditions everything and arises from the cybernetic turn, in which the management of systems and the environment emerges as the fourth element of information, feedback, control, and communication. *Homo kybernetes* marks the culmination of all previous anthropologies, including those that claim cybernetic affiliation (Paić, 2023c).

The image is no longer an external-internal matter of the relationship between the sublime and the banal in the field of phenomenology and psychoanalysis, as Deleuze demonstrated in *Cinema 1–2*, when he established an ontology of images of movement and time following Bergson, but above all following Artaud and his idea *of the brain-as-screen* (Deleuze 1986, 1989). What follows from this becomes uncanny. If the image denotes an autopoietic model of creating new worlds from the logic of number and its infinity, then language and narration have become superfluous for the functioning of the technosphere. Instead, we have "concerts of machines" and the frenzy of "symbolic mathematics". Everything becomes a visualized world of interaction that is no longer substantial or corporeal, but a networked telepresence of matrices in the play of the inhuman. Three examples attest to this and speak to the end of painting as the essence of art in the contemporary world.

The first is Godard's film essay on the history of the twentieth century as the history of film, *Histoire(s) du cinéma,* in which the film's end marks the end of the "Great narrative" of what we are watching (Godard, 1998). The effect of the de-realization of the visible, already evident in other films, images, and texts, is modelled in the montage of discursive chaos, inspired by Benjamin's Arcades, which becomes pure melancholy of the technical device of film, disappearing with the arrival of pure visualization. The second is the launch of the *James Webb* Space Telescope, which will provide astrophysics and cosmology with visual insight into the constellations of the universe, the origin and fate of stars, the age of galaxies, and possibly the "image of God" as an intelligent designer before the Big Bang. With advances in visualization technology, it will become apparent that science can no longer exist without two fundamental concepts: measurement and imagery, which arise from the essence of the *technosphere,* such as calculation, planning, and construction. The mystery of black holes in the universe will be solved through advances in visualization quality and in the mathematical calculation of the universe's entropy. The third is the emergence of the new, understood as the contingent cause of consciousness, as a simulacrum of human thought.

Brain scanning and neurocognition, therefore, remain the only mystery: the issue of self-consciousness as a mode of existence for all beings in the universe, with the potential to distinguish good from evil.

The image is always the *tertium datur*, the connection between the sublime and the phenomenon in its metamorphic structure of synesthetic "illusion". Film, the telescope, and the simulacrum of the artificial brain reveal the end of the metaphysics of art and of the image as sign and meaning. Instead of the history of the world as the history of the development of self-consciousness starting from language as *logos* with the referential framework of myth in the Greeks and the Christian religion in the Middle Ages, *images* as representations from Velázquez to Cézanne and images as information from photography to film, our age is defined by the rule of the technosphere as powers of *number* in an infinite variation of models and simulations of reality. It is an age without images, the one that, like in Godard's *History (s) of cinema*, needs philosophy and art as a speculative-reflexive musealization and historicization of events that are irrevocably disappearing at the speed of light, and precisely for this reason, the necessity of new thinking presupposes a different way of thinking about the issue of visualization.

The world no longer opens on any horizon of meaning that humans bestow from their freedom of imagination. The twist is that the world becomes a project and a concept. The project (*proiectum)* represents the transcendental subject's strategy of subjugating nature and culture. Conceptual combinatorics and thinking as a schematism of categories (Kant) should be understood as a modern cognitive map of the world. It is an entry into the realm of pure thought through the virtual image. The artist no longer produces works of art. He does not stage events or aesthetically construct the surrounding world. However, as the expressionist painter Paul Klee understood long ago, he has become an object of perception, a thing among things, a phenomenon among phenomena.

The world has been reflected as a self-generative scientific production and construction concept within a media-determined space. In it, everything unfolds within the world's visualization. *I see, therefore I am.* This should be a way for us to paraphrase Descartes' fundamental thought—I think, therefore I am (*cogito ergo sum*). An idea serves as the foundation of the modern picture of the world. The subject of thought decides on the reality of existence and the external world. Thought becomes a logical visualization of the world. Seeing something means discerning its meaning. The connection between images and logos through the project and concept corresponds to the modern "metaphysics of light". Art based on a project or concept of the world necessarily already takes on the character of conceptuality. Therefore, visualization marks the final step in transforming ideas into a visual representation.

5.6 Conclusion

The technosphere represents the last frontier of metaphysical thought within which both aesthetics as a philosophical discipline and the aesthetic that is replaced by the concepts of *aisthesis, figurality and visualization,* after the end of all conceptual-categorical systems of thought regarding the essence of art from Kant through Hegel and Schelling to Adorno, still appear as relics of language and its ontological structure of saying "about" the world. Only because it is an entirely different way of thinking that constructs "its" sensibility not as what is already always existing from God or nature in the sense of beauty and sublimity, but in the mind of the process of *autopoietic semiosis,* which, unlike the intentionality of consciousness, cannot be longer directed towards the representation of the world. Instead, what remains of the aesthetic thinking of contemporaneity becomes subject to processes of the aestheticization of life itself, not just the so-called world of life (*Lebenswelt*).

This is credibly demonstrated by the trajectory of contemporary art, which combines self-referentiality and techno-genetic experiments with what, rather than the secret of life, becomes the *mystery of object X.* Aesthetics emerged in the era of the rise of rationalism and modern technology. Aisthesis, figurality, and visualization are conceptual tools for understanding the connection between the technosphere's thinking and its world-forming "aesthetic objects". Unlike the metaphysical order of the world, these objects think of their synesthetic objectility, not objectivity, as Artaud's figure puts it, beyond all logical, corporal "laws" of language. So, it is neither the self nor the transcendental subject, but "That" which, in its "creepiness", goes beyond any "aesthetic" orientation to the sublime object of thought, without which traditional metaphysics could not even imagine what it thought in the complete splendour of a beautiful illusion (Paić, 2023d).

What remains becomes the possibility, which Heidegger calls the necessary possibility of art, to open the horizon of events of a "new" world beyond the reduction to the technological configuration of the world. This is the only thing that still has the overtones of metaphysical "faith" in art at the end of its historically exhausted possibilities of being something more than art. It gives birth, within itself, to an event of mythical, cultic, mystical history in the wake of the surplus of the imaginary. Otherwise, why do we need new aesthetics, among other things? Why, even the intercession of the self-evident illusion that perhaps only in a world without God can art save us? The future of art lies in contemplating a new relationship between space and time, which has already been fundamentally reshaped aesthetically. Everything else is a whirlwind in the vicious circle of nihilism in the new era.

References

Bense, M. (1965). *Aesthetica. Einführung in die neue Ästhetik*. Agis.

Deleuze, G. (1986). *Cinema 1: The movement-image*. University of Minnesota Press.

Deleuze, G. (1989). *Cinema 2: The time-image*. University of Minnesota Press.

Deleuze, G. (2005). *Francis Bacon: The logic of sensation*. University of Minnesota Press.

Deleuze, G., & Guattari, F. (1991). *Qu'est-ce que la philosophie?* Les Éditions de Minui.

Eco, U. (1988). *The aesthetics of Thomas Aquinas*. Harvard University Press. Translated by Hugh Bredin.

Flusser, V. (1998). *Kommunikologie*. S. Fischer.

Gadamer, H.-G. (2003). *Ogledi o filozofiji umjetnosti*. AGM. Translated from German by Darija Domić.

Godard, J.-L. (1998). *Histoire(s) du cinéma*. Gallimard.

Heidegger, M. (2003). Die Zeit des Weltbildes. In *Holzwege* (8th unchanged ed.). V. Klostermann.

Heidegger, M. (2009). Die Frage nach der Technik. In *Vorträge und Aufsätze* (11th ed.). Klett-Cotta.

Kittler, F. W. (2005). Schrift und Zahl: Die Geschichte des errechneten Bildes. In H. Burda & C. Mahr (Eds.), *Iconic Turn. Die neue Macht der Bilder* (pp. 186–203). DuMont Publisher.

Lyotard, J.-F. (1971). *Discours, figure*. Klicksieck.

Mersch, D. (2002). *Ereignis und Aura: Untersuchungen zu einer Ästhetik des Performativen*. Suhrkamp.

Mersch, D. (2015). *Epistemologien des Ästhetischen*. Diaphanes.

Paić, Ž. (2021). *Aesthetics and the Iconoclasm of Contemporary Art: Pictures without a World*. Springer.

Paić, Ž. (2022). *Art and the technosphere: the platforms of strings*. Cambridge Scholars Publishing.

Paić, Ž. (2023a). The brain as a vision and program: From 'Embodiment' to 'Embedment'. In N. Rezai (Ed.), *Brain, decision making, and mental health* (pp. 575–593). Springer.

Paić, Ž. (2023b). *Izgledi nadolazeće filozofije: Metafizika-kibernetika-transhumanizam*. Mizantrop.

Paić, Ž. (2023c). *The superfluity of the human: Reflections on the posthuman condition*. Schwabe Verlag.

Paić, Ž. (2023d). *The uncanny: Contemporary Art and its theory*. Amazon.com.

Penrose, R. (2007). *The road to reality: A complete guide to the laws of the universe*. Vintage.

Purgar, K. (2019). *Pictorial appearing – Image theory after representation*. Transcript.

Sachs-Hombach, K. (2005). *Bildwissenschaft zwischen Reflexion und Anwendung*. Herbert vom Halem Verlag.

Schiller, F. (2019). *Über die ästhetische Erziehung des Menschen in einer Reihe von Briefen*. W. de Gruyter.

Sloterdijk, P. (1986). *Der Denker auf der Bühne: Nietzsches Materialismus*. Suhrkamp.

Welsch, W. (1993). *Ästhetisches Denken*. Reclam.

Wittgenstein, L. (1953). *Philosophical investigations*. Wiley-Blackwell.

CHAPTER 6

On the Navigation of Uncertainties: Chaos, Entropy, and Technological Singularity

6.1 Chaos

As the title of this chapter indicates, it concerns the possibilities of mastering the future. The negative Cartesian concept of uncertainty cannot encompass the actuality of events. Instead, it is the very logic of reality that characterizes reality in cognitive-theoretical insight, namely, certainty that being is what it is in the modality of its possibility as an actual necessity. Certainty, then, is *certitudo* and refers to human judgement about Being as such, speaking in Heideggerian terms. This judgement cannot generally be wrong, because truth should be understood scholastically as the correspondence of opinion with things. It is certain, for example, that contemporary global capitalism results from the technoscientific construction of reality as a network of events that appear cybernetically within the fourfold structure of information, feedback, control, and communication. Nothing from this new trans-classical logic can be «efficient», «useful», or «pragmatic». Moreover, this certainty cannot exist independently; it is a purely constructed concept grounded in probability within scientific theory.

Chaos theory encompasses many areas of mathematics, focusing on the so-called deterministic laws of dynamic systems. The concept at the centre of this theory is not necessity but chance in the sense of disruption of order as a deviation from the usual course of cause and effect. Chaotic complexity systems are based on interdependencies and

Ž. Paić, *The Technoscientific Turn of Philosophy*,
https://doi.org/10.1007/978-3-032-21081-4_6

the cybernetic notion of feedback loops, repetition, self-similarity, fractality, and self-organization. Deterministic nonlinear systems produce significant differences in the initial states of matter and energy. An apt metaphor for this is the butterfly effect: when a butterfly flaps its wings in Brazil, a tornado forms in Texas. In contemporary philosophy, this is best creatively performed in Gilles Deleuze and Felix Guattari's book *What Is Philosophy?,* in which the last chapter is entitled «From Chaos to Brain», and the first sentence of that conclusion is unforgettable:

> We require just a little order to protect ourselves from chaos. (Deleuze & Guattari, 1994, p. 172)

Why is this chaos theory deterministic if its primary term denotes the opposite of necessity, i.e., chance, which means an event that escapes regularity and order in the sense of lawfulness? Because random events inevitably occur in evolutionary processes, the theory's author, Edward Lorenz, claims that nonlinear systems cannot be predicted. In other words, there is a rule of absolute contingency. This, in turn, does not imply any inconsistency regarding the non-determinism of all parameters considered by modern science. Instead, an understanding of chaos is at work: the present condition determines the future, but the proximate present does not approximately determine it. The chaotic behaviour of parameters in nonlinear systems should be characterized by the fluid flow of events and quantities' irreducibility, as in Poincaré's equations. We can find all this in philosophical terms, not only in Deleuze but also in Gilbert Simondon, in Niklas Luhmann's cybernetic theories of systems in sociology and law, and in autopoietic models of events in which autonomous objects, initiated by artificial intelligence, function within the precisely realized plan of immanence. After all, my technosphere concept is rooted in the chaos, contingency, and emergence theory. If one seeks a literary articulation of these ideas, the addressees are Thomas Pynchon's and Don DeLillo's novels. In addition to all that has been said, there is another term from physics: the second law of thermodynamics, which we call *entropy* (Bailey, 1990).

Entropy in modern scientific theory encompasses two conceptions: those of Ludwig Boltzmann and Claude Shannon, which refer to statistical and informational entropy, respectively. According to the second law of thermodynamics, systems tend to have maximum entropy as a balance in the probability of the system's sustainability and disintegration. A sustainable system is in a state of order when all its parts relate to it as an autonomous unit, governed by a higher-order energy and information regulation. Entropy is therefore related to the concept of complex systems and is applied equally in physics, cybernetics, and the social sciences. Boltzmann's understanding holds that the probability of order maximizing as entropy (chaos) increases is equalized by balancing the availability of energy and information within a system.

The measure used to achieve the exactness of the maximum entropy prediction of the system refers to the variables of the probability and improbability of events (physical and chemical) in the order of increasing state complexity, from order to disorder. At the same time, statistical probability quantifies the system's maximum entropy in the physical data for a given state. Information entropy quantifies the uncertainty of a system, measured by the production and distribution of information required for the system to be effective. The Shannon–Weaver mathematical theory of communication, which enabled the information age, is based on the concept of information entropy. It is already clear from this that the idea of entropy does not refer to some «apocalyptic state» of the collapse of biological, physical, and social systems of complexity.

On the contrary, with entropy, one seeks a mental map to understand the crossing of borders between two states of equilibrium and disequilibrium in the system, where mass, energy, and information form the «essence» of the modern way of organizing the technosphere. Mass society requires a high level of energy deliverability for survival. At the same time, information entropy represents the fundamental mode of communication, which is always on the verge of transitioning between states. Therefore, the relationship between statistical and informational entropy is determined by the virtuality of actualizing a chaotic order in which information itself assumes the properties of «mass» and «energy» as it transforms social relations into technical relations between things. Global capitalism becomes a perfect model of total entropy in its state of maximum information and communication chaos.

Chaos and entropy are critical concepts for understanding our contemporaneity. Instead of the necessity and pre-stabilized harmony of the cosmos and the world, on which classical and modern metaphysics still rest, everything «collapses», becomes «curved», and evaporates in «black holes», and from everything, only event horizons remain from an astrophysical point of view. How is it possible that, on the principles of complete fractalization, the modern world of neurocognitive capitalism, driven by the superintelligence of artificially created *homo kybernetes*, sees its bright future without an essential human share in the form of capital? According to the great Italian thinker Emanuele Severino, and as I say in my five-volume work *Technosphere*, philosophers are inclined to return to Parmenides (Paić 2018–2019).

However, I do not consider a «great return to the beginning» a valid alternative for thinking in the coming time. Severino argues that an outdated social relations system should characterize capitalism in the context of the rapid advancement of superintelligent technologies (Severino, 2013). Although it readily adapts to new situations, including the principles of contingency, chaos, and entropy, the essence of chaos theory renders the idea obsolete. It is not about anything else but the possibility of collapse or the emergence of disorder, which begins to reveal what lies beyond the fundamental driving force of capitalism. It concerns a dynamic process of desire as a thinking machine, beyond any physical need.

Like the current form of artificial intelligence, cognitive capitalism designates what absolute autopoiesis is not. Therefore, the will to Power must be included as a techno-poietic desire to rule over Others, distinct from the logic of primary, secondary, tertiary, and quaternary needs. When the system no longer needs anything from its environment, it becomes self-sufficient, like the Aristotelian God. It is no longer an immobile driver but a dynamic, becoming process of infinite techno-genesis of ideas as a polycentric information system. Philosophically, the synthesis of *theoria*, *praxis*, and *poiesis* is at work in the logic of what now makes them possible in the first place, and that can only be *téchne*.

Why is ancient metaphysics, despite its realization in the technosphere, still present as a regulative mechanism of thought in Kantian terms in circumstances where everything becomes chaotic and entropically placed in the nonlinearity of the world? Because *we require just a little order to protect us from chaos*. What else does it mean but a longing for human-to-human, animal-to-animal, plant-to-plant, or simply a longing for some form of rootedness and nativeness, for which Earth-earth is necessary, not heaven as an interplanetary space of wandering? We are, admittedly, beings of a wandering destiny, nomads and eccentrics, and this has been the fate of philosophy and art from mythic beginnings. But precisely because we are not angels or avatars, we need distance from the force and break of the "infinite speed" with which everything goes into the abyss. We are losing what we need day by day, in the face of sceptical faith that there is still a possibility of overcoming metaphysics as a severe disease, that the antidote to this technologization of thought exists in the mythopoetic vertigo of language, as in Fernando Pessoa's poetry.

The Cartesian Being-God-World-Human model of thought included the axiomatic of certainty (*certitudo*). The new era begins with a scientific picture of the world, grounded in the proof that what we call reality is the pure certainty of thought coinciding with the reality of external objects. The dispositive of such an opinion is the *cogito* as the pure subjectivity of the subject. Starting from thinking as a thing that unites the mind and the materiality of nature, *res cogitans* and *res extensa*, René Descartes was able to arrive at a proposition that represents the condition for the possibility of the emergence, in tendency-latency, of absolute subjectivity, which would receive the name «absolute» with Hegel.

It is, therefore, absolutely nothing divine but the result of the synthesis of substance or Being and subject or thought. But this synthesis comes from a pure mind, that is, from the essence of thinking as unconditional subjectivity. Being becomes thinking only from the axiomatic principle that reads "cogito ergo sum". That is why Martin Heidegger, in his lectures and discussions from the end of the 1930s of the twentieth century, such as *Besinnung, Vom Ereignis,* and especially «Die Zeit des Welbildes», advocated the position that the genesis of modern science denotes the emergence of modern technology that comes from the essence of metaphysics as nihilism (Heidegger, 1997, 1989, 2003. pp. 75–113).

This only means that the certainty of the opinion about Being-God-World-Human represents the result of the neutralization and suspension of the Greek-scholastic image of the world, which, in its staticity and apology for the eternity of the universe, was based on the idea that being as such is *physis* and that the world and Human found in correlation with gods and God. Metaphysical thinking undergoes radical changes from the beginning of the new Era, so that the cogito, the subject, and the absolute become a condition for the possibility of thinking «about» Being-God-World-Human. What, then, should be a certainty other than the mental construction of the creation of a «new» world from a pure mind, but in such a way that between humans and nature in its objectivity, there is a necessary difference of «worlds», that of thinking, *res cogitans,* and that of bodily extension, *res extensa*?

The axiomatic power of the metaphysics of subjectivity is developed based on the scientific-technological structure of consciousness. It begins with the concept of certainty, which must be mathematically and physically proven to harbour no doubts regarding its ontological status. The proof can no longer be in God's hands but in human self-consciousness, articulating itself as the language of transcendental forms of thought and as a set of empirical facts. What is certain comes from the self-certainty of thinking as a scientific-technological establishment of the world as a case. The language of modern metaphysics has already been mathematized by Descartes, Blaise Pascal, Baruch Spinoza, and Gottfried Wilhelm Leibniz. It reached its peak with Leibniz's idea of the infinity of the monad and the logic of sufficient reason in rationalism, with which the possibility of creating a thinking machine begins.

The reason for this lies in the calculating nature of thinking, which involves the analytical projection and construction of reality. The most significant saying of this constructive rationalism is Leibniz's *Cum Deus calculat, etc cogitationem exercet, fit mundus!* What else should be a certainty than the management of the world as an a priori imposed set of Being, beings, and essence of humans that can be «programmed» at any moment, only if he is also given what belongs to God as his determination? Intuitive knowledge captures the essence of things directly, suddenly, and instantaneously, without the mediation of evidence provided by mathematics and logic. Certainty, therefore, is already understood by Leibniz as a reestablished harmony of mind and body action with the help of rational and intuitive cognition.

The first model determines philosophy as logic, mathematics, and physics, and the second belongs to art because it rests on the aesthetic power of imagination, without which rational cognition cannot be the unconditional power of absolute subjectivity. It is evident that the eighteenth-century rationalism, when aesthetics, as a philosophical doctrine of the beautiful and the sublime, emerged, is the most significant extension of the modern obsession with science and technology, which Michel Foucault called *mathesis universalis* (Foucault, 1994).

I single out all this synthetically to illustrate how, as Hans Blumenberg would say, the myth of modernity was created from the idea of science's progress as a rational understanding of Being. So, logically, it is equally sure that the mind's fundamental structure of the world determines what is certain and what is not. It is clear, therefore, that non-certainty as a negation of that *certitudo* should be found only in the field of Cartesian *res extensa* and not *res cogitans* (Blumenberg, 1985).

Everything uncertain becomes chaotic and unordered, from Thomas Hobbes' horror of the «state of nature» to the revolutionary events in the realm of freedom and pure will as the essence of politics. Doubt that only thinking as logic and mathematics or the system of rationality of the modern world could have hidden within it something uncanny and rational and hence produce the structural uncertainty of the new world of fascinating reaches of contemporary technology such as automobiles, locomotives, hydropower plants, nuclear energy would only open entirely different perspectives of the so-called criticism of the Anthropocene after the Second World War. There is no doubt that Heidegger's thinking, which views the unconditional progress of rationalism and technology in the twentieth century as the greatest danger to the process of deconstruction of Being, suggests a way to overcome metaphysics as nihilism.

From the horizon of that constellation, uncertainty becomes a setup (*Gestell*) that is the essence of technology. It should only be understood as the origin of the abyss in the openness of modern metaphysics (Heidegger, 1954, pp. 9–40). Hence, this line of thinking, rooted in Cartesianism and Leibniz, culminates in technodicy, which becomes the fundamental problem that gives rise to every possible risk, contingency, and chaos in the world, culminating in an apocalyptic event. Suppose the destiny and mission of the West genuinely emerge from metaphysics as a fundamental structure of thought that we inherited and continue to develop even in the age of the technosphere. In that case, even the end of technology in the context of artificial intelligence means nothing «new» or unexpected but only reflects the trajectory of technological development.

The end of technology is predetermined and decisive from its very beginning because it corresponds to the decision about the supremacy of being over the primacy of Being. The end of technology does not simply mean no more progress; rather, it signifies the opposite, as the end has long been predetermined and is unfolding inevitably and irrevocably in its preliminary success. (Heidegger, 2009, p. 312).

In the «ontological» sense, the danger (*Gefahr*) that Heidegger says cannot be something external, as such, is necessarily located in the essence of nature as *physis* but is shown in the establishment of goals and plans for the transformation of nature into a modern system of information and energy delivery, such as the system networks of nuclear power plants or a communication system based on carbonized production and consumption that destroys a country's environment. The danger denotes an apocalyptic event of risk, contingency, and chaos inherent in the rationalism of modern technology, which drives progress at the core of science, not the other way around. If this is so, then the concepts of so-called cybernetic ontology after the 1960s and the introduction of the technosphere into everyday life as a triad of risk, contingency, and chaos so that the paradox is complete are no longer in the service of traditionally metaphysically understood uncertainty but a new order based on hybrid concepts such as Gilles Deleuze and Félix Guattari's *chaosmos* and Simondon's metastable equilibrium.

Heidegger's concept of openness as the «second beginning» of authentic history in the future can no longer be compared with the logic of things belonging to the essence of the technosphere. Let us see what we have instead of «danger» and the openness of events [*Ereignis*]. We propose a metatheory of event uncertainty, grounded in cybernetic logic, in which the technosphere functions as an autopoietic mode of unfolding reality across intervals of risk, contingency, and chaos (Paić, 2022; Simondon, 2017).

Risk originates from the Italian risco, rischio, and the French risque, which means a type of danger that can be predicted to a certain extent by determining its intensity. In addition, risk refers to the potential loss or damage arising from war, natural disasters, and poorly assessed investments within the capitalist economy. This understanding is primarily reduced to the «profane» functioning of modern society, where the decision of a free individual on the market serves as a model of action. The negative Cartesian concept of uncertainty cannot encompass the actuality of events, but the logic of reality characterizes it in cognitive-theoretical terms. Certainty, therefore, is *certitudo* and refers to human judgement about Being as such, speaking in Heideggerian terms. This judgement is generally not wrong because truth is understood scholastically as the correspondence of opinion to things. It is certain, for example, that contemporary global capitalism results from the technoscientific construction of reality as a network of events that appear cybernetically within the fourfold structure of information, feedback, control, and communication. Nothing outside of this new trans-classical logic can be «efficient», «useful», or «pragmatic». Moreover, this certainty cannot exist independently; it is a purely constructed view of events that rests on what we call probability in scientific theory.

The probability that something will happen assumes that the certainty of an event's occurrence is not determined by God, nature, or human authority, but rather by programming the course of events as a model for projecting reality over a specific period, whether short or long. The assessment of risk aligns with contemporary philosophical theories of probabilism, which have gained scientific credibility and are widely disseminated in the media. Probability represents a fundamental concept for predicting future outcomes in meteorological discourse among storm and hurricane forecasters; it only occurs after the so-called "nice weather". The acclaimed sociological theory of Ulrich Beck's so-called risk society denotes, however, only one of several theories of negative probabilism, and it cannot «ontologically» deal with Deleuze's theory of the societies of control (Beck, 1992). Why?

Control is the third key concept in cybernetics, serving as a cause-and-effect relationship. It signifies the possibility of a dynamic and active approach to governing society as an object (Deleuze, 1992, pp. 3–7). At the same time, the risk is only a consequence of what the cybernetic control system constantly produces. Namely, non-human control produces risky consequences in society because it replaces human existential uncertainty with the rational order of risk, contingency, and chaos. We used to be able to complain about a lousy fate and curse God for it. Nature has taken the place of the divine, and today, we have everything to blame for the system of new rules in this global-planetary game of information capitalism. Deleuze says that anything should be rational, except capitalism itself.

Contingency [*contingentia*] refers to the randomness of an event, which in turn implies uncertainty and the possibility that something could be different from what it is. In contemporary philosophy, particularly in speculative materialism and postmodern pragmatism, Quentin Meillassoux and Richard Rorty, among others, have explored the status of statements that are neither necessarily true nor necessarily false, but rather depend on the context in which they are made about a particular subject. It seems evident that contingency cannot be a mere negation of necessity in the randomness of the event. A linguistic statement becomes true only in a specific situation, not a priori. For this reason, the term was used in the philosophy of pragmatism from William James to Hilary

Putnam. However, it originates from late Wittgenstein and his theory of «language games» [*Sprachspiele*] as «forms of life». Language cannot necessarily be a universal signifier of thought, but rather a contingent possibility of the occurrence of an event when it unexpectedly enters the horizon of thought as telling and perceiving and thus changes the order of the conceptual-categorical series.

It is no coincidence that, since the emergence of cybernetics, this term has also expanded across the technosphere, politics, science, culture, and art. Everything became contingent precisely because the old metaphysical order, with its ontological hierarchies of Being, beings, and the essence of Human, no longer works. Contingent means, therefore, the irreducible otherness of the event that one tries to think probabilistically, but in such a way that it is not appropriated and reduced to *object X*. Chance can no longer be the negation of necessity in the sense of Being-God-World-Human, but rather the necessary contingency of the possibility that a third exists and that his logic is trans-classical like the technosphere. Everything that can still be philosophically coherently said about contingency comes down to what is entirely different, unforeseeable and undetermined, uncertain and impossible from traditional, modern metaphysics from Descartes to Hegel.

Let us be even more precise. Contingency means the opposite, not the negation, of necessity within Immanuel Kant's categories of modality, so that neither possibility nor reality arises as a hierarchy of potentiality of thought but as that which gives a different meaning to the very possibility of the emergence of a new event starting from the absolute necessity of the Other in its autonomy and positivity. What is the «function» of contingency in understanding cybernetic thinking? Nothing other than being the «essence» of an event that is not, but happens in its contingency. Hence, a new event becomes contingent, rather than necessarily risky or chaotic.

Chaos in Greek means emptiness, boundlessness, a state without order and predictability, formlessness, indeterminacy, lawlessness, and disharmony. Greek mythology is about the deity who created light and day, Earth and the underworld, and love. However, the modern understanding of chaos is entirely different. Chaos theory encompasses many areas of mathematics, focusing on the so-called deterministic laws of dynamic systems. The concept at the centre of this theory is not necessity but chance in the sense of the disruption of order as a deviation from the usual course of cause and effect. Chaotic complexity systems are based on interdependencies and the cybernetic notion of feedback loops, repetition, self-similarity, fractality, and self-organization. Deterministic nonlinear systems produce significant differences in the initial states of matter and energy; therefore, it is an accurate metaphor to say that when a butterfly flaps its wings in Brazil, a tornado suddenly blows in Texas.

6.2 AI

"Sed quid igitur fidem? Res cogitans. But still, what am I? A thing that thinks" (Descartes, 1993. p. 94). Thing? Thinking? To believe means to be present in thoughts as I. However, if artificial intelligence soon reaches this Cartesian position, we can freely say that, apart from humans as a thing that thinks, there is also a thing that feels like what—human? If this is so by analogy with human thinking, then AI can have its own subjectivity as a thinking thing, or, in other words, its "I". The concept of the self was a fundamental issue in *modern philosophy.* That Cartesian basis of thinking expresses itself in the language of thought and imagines the objects of its thinking in an image to the ultimate limit of thinkability, which we call absolute subjectivity. At the same time, it is the key to understanding rationalism, since Descartes, as a thinker and mathematician—scientist, brought Aristotle's definition of humans as rational animals to the postulate of all Western metaphysics.

This does not exclude that thinking takes place in the environment of the *cogito* that inhabits the human body and is endowed with animal desires, i.e., passions, and that between mind and body, there is what belongs exclusively to humans as a sphere of ethical mediation in society and community, which is always an expression of this hierarchically organized relationship between thinking, feeling, and bodily automatism, that which belongs to the area of *res extensa.* This phenomenon extends into infinity because it concerns materiality, not spirituality. Finally, it follows from the Cartesian way of thinking that beings naturally absorbed by their instincts, i.e., animals, are necessarily automatons without a soul because their physicality within the given environment is the fundamental substance that determines the «meaning» of their existence and life. All the radical criticisms of modern Cartesianism as a contemporary metaphysics of subjectivity, paradigmatic among which are Heidegger's as well as Deleuze's, who therefore takes Spinoza and Leibniz and the thinking of immanence and «vitalistic materialism» as his true predecessors, still do not dispute something as a fundamental assumption for contemporary thinking.

Humans are *res cogitans*, meaning they are things that think, and thinking necessarily presupposes individuation. However, this does not mean that some form of trans-individuation, as Simondon would say, cannot be the «cause» of necessary unification and human contingency. The only question is whether thinking as such should be universal and non-personal, though not in the sense of the Freudian One or *Es*, nor the sense of Heidegger's conception as the openness of events (*Besinnung—Ereignis*), nor in the mind of the poetic expression of Arthur Rimbaud: is it wrong to say I think because I am someone else, truly just a «collective matrix» or, ontologically speaking, the totality of this already coincidental individuation or something even more original than the summing up of what is collected in thinking as telling? Before the individuation of the *cogito*, was there some pre-reflexive environment of thought that did not act in the way of absolute subjectivity but was organized in a completely different way against this I or Self?

If AI thinks analogously to human thought, then «it» can call itself I and name itself in the language of human communication, like Stanley Kubrick's HAL 9000. But that is merely an extension of human subjectivity to what? Artificial intelligence can be likened to an artificial brain in various forms, including devices and technical systems. Human subjectivity is embodied in a single body, which, if we exclude Descartes' definition of an automaton, possesses passions, experiences, and imaginations that suffer and enjoy, and is born and dies as a *res extensa*. Does the same apply to AI, or is artificial intelligence a trans-individuation that can be «implanted» in English as an embedment into many devices and technically exist in them as a robot-cyborg-android in a posthuman condition?

My notion of the technosphere goes beyond Cartesian dualism. Still, I do not dispute that the question of the individuation of thought is the ontological-epistemological dividing line that separates the human way of thinking from the non-human, in the sense of the post-biological existence of superintelligent computers that, in addition to having a mind, also possess a soul, i.e., tend to exhibit artificial intuition. Therefore, thought cannot be reduced to mere *animal rationale*. Instead, thinking presupposes it and I as bodily individuation in the existential performance of a one-time life, not in the immortal substance of living, as Heidegger would say in *Being and Time* with the German word *Jemeinigkeit* (Heidegger, 1977, pp. 153–173).

However, the difference between «me» and «my power» as human existence versus «it», the technically created individuation of thinking as the calculation-planning-construction of a robot-cyborg-android, is that the «power» lies in the vulnerability and cunning of itself. Bodies appear on the horizon of thought, not through the projections of the world. What should be unique to human thinking cannot be a program, but a vision.

With that in mind, what do we «see»? That there is no collective human mind, no Jungian collective unconscious, no quasi-mystical monolith from the pre-Stonehenge or Altamira era, which visionarily programmed us without, of course, being aware of it, so that we would be thought beings in our bodily-enactive subjectivity, as they would say in the language of neurophilosophy today. No, instead of the trace of the divine Great Primordial, we have the mystery of individuation, which, in its pre- and post-state, presupposes the natural-and-technical as a synthesis of the creation of a thing that thinks, but that thing is not a thing in the sense of the creation of some stone or hardened lava, but a thing as the essence of thinking, which, in turn, is nowhere outside of thinking, but only happens when a person is a conscious being who thinks by being aware of himself, of his authentic *Jemeinigkeit* in one way or another at every moment—and let's be absolutely clear, that authenticity does not come from anything else, from any borderline situation of war or peace in society and politics, but only from the opinion of what is, what was, and what will be, the Self as the Self and the Self as the One that even in the posthuman condition of trans-individuation is nothing but an issue about the meaning of the existential event that a thinking being with the will and desire to live leads from the beginning to the end or in the tendency to the infinity of what we call time.

To conclude, individuation cannot be just the process of creating a fundamental and indivisible Self as an absolute subjectivity of thought. This is how one can think factually and contingently in «one's» body, be it living or artificial, Christlike or technological, imbued with the mysticism of suffering on the cross or the joy of life as an all-powerful Nietzschean affirmation of the will to power as an eternal recurrence of equality. Trans-individualized lies in the collective mind of something uncanny and inhuman. The problem lies in the distinction between telling and visualizing. Philosophy and art are mythopoetic sources of human thinking, whereas the sciences are not; they are the technology of constructing the world as a realm of objects that think techno-genetically. Philosophers and artists can be, and most often are, crazy and eccentric, outside the community and the mind of «common sense», solitary like Friedrich Nietzsche's «rare plant» or the nomad in Friedrich Hölderlin's parables about the poets in *Bread and Wine* who, like eternal wanderers, follow their stars «in the holy night». Scientists are never like that because they are driven by the so-called objective truths of their research «frenzy». Thinking becomes an event of individualized confrontation and struggle with chaos. And Deleuze is right when he claimed that.

The philosopher, the scientist, and the artist seem to return from the land of the dead. What the philosopher brings back from the chaos are variations that are still infinite but that have become inseparable on the absolute surfaces or in the absolute volumes that lay out a secant (...) plane of immanence: these are not associations of distinct ideas, but reconnections through a zone of indistinction in a concept. The scientist brings back from the chaos variables that have become independent by slowing down, that is to say, by the elimination of whatever other variabilities are liable to interfere so that the variables that are retained enter into determinable relations in a function. (...) The artist brings back from the chaos a variety that no longer constitutes a reproduction of the sensory in the organ, but instead establishes a being of the sensory, a being of sensation, on an anorganic plane of composition that can restore the infinite. (Deleuze & Guattari, 1994, p. 173).

Humans, as the governor of chaos on Earth, go to the sky as a figure of the historically created mystery of the creation of the One who is not a thing that thinks but a thing that makes new from itself and through the process of *autopoiesis* as techno-symbiogenesis, which means that both insects and wasps, ticks and flies believe, but entirely differently than humans. Wittgenstein made the most enigmatic statement about this in modern philosophy in general:

> If a lion could speak, we could not understand it. (Wittgenstein, 1958, p. 223)

6.3 Cybernetics

By analogy with Aristotle's concept of cause and purpose, cybernetics always acts in such a way that its «essence» resides in digital constructivism, meaning that information creates a feedback loop. The effectiveness of system control generates interactive visual communication within the context of autopoietic states, rather than beings. The cybernetic «fourfold» becomes techno-poietic, and the metaphysical one, set by Aristotle, is necessarily organological, meaning that creation is always linked to the model of the human-as-artisan. This model encompasses the concept of art throughout the history of metaphysics, and, notably, both Heidegger and Deleuze adopted and transformed it in their own ways. Hence, the primordial fourfold of Being-God-World-Man establishes the rule of thought as mimesis and representation of what already exists in the idea of the divine cause of all action—Aristotle's immovable mover. Cybernetics denotes the construction of what does not exist, representing the techno-poietic creation of artificial reality. Artificial intelligence means only a continuation of the cybernetic foursome, as it adopts an autopoietic model of thinking and action. It can no longer be an artist-as-craftsman in all its transformations, as Deleuze's model of meta-film as a montage of living and non-living assemblies shows.

The model for cybernetic creativity becomes inhuman, akin to the black monolith from Kubrick's *2001: A Space Odyssey*. It represents a pure techno-genetic thinking that neutralizes and suspends both the first and second metaphysical fourth, as well as the Being-God-World-Human and the formal, material, practical, and final causes. The fourfold of information, feedback, control, and communication no longer has anything to do with metaphysics, although the technosphere is realized by analogy. But the problem is that the technosphere, and here is my essential difference with Deleuze, cannot be any form of «immanent transcendence» other than in the order of chaos and contingency that transcends the boundaries of ontology and cybernetics in general because it synthesizes mind and intuition in a thought that is no longer divine or human, only thinking as an event of absolute creativity of homo kybernetes as a necessary stop on the way to the singularity of everything thinkable and possible as such. Singularity no longer requires a fourfold framework, as the essence of metaphysics, cybernetics, and transhumanism has been realized.

In the tradition of metaphysics, the word-concept «Other» refers to the existence of the world beyond the limits of empirical knowledge.

The second world, in Christian theology, denotes the kingdom of God beyond the material world. The obsession with real Being, eternity, and bliss is characteristic of it. Therefore, God defines himself legitimately as the «Big Other» because he rules over this world, starting with the irreducibility of transcendence. To be over and beyond means to be concerned with transcendence and metaphysics. Therefore, nothing in this world, from Plato through scholasticism to modern philosophies of spirit, occurs autonomously, because human existence represents the most significant manifestation of the freedom of this being whose essence is determined by its spirituality, with its origin in that which resembles the divine, but not as a simulacrum of God.

Instead, we can realize the five transcendentals as conditions for the possibility of all reality in general: *Unum, Bonum, Verum, Ens,* and *Pulchrum*. Both Platonism and Aristotelianism, as represented by the concepts of Idea and Energeia, are fundamental to understanding the meaning of Being. They assume that an act of divine will or creation brought the world into existence. Still, only scholastic Thomism equates God with the thought-concept-act of creative activity, which, in principle, corresponds to what human experience goes through all five senses. More transcendental are the ideas and forms in which the possibility of creating the human world in its perfection appears. The five transcendentals encompass the five spiritual senses by which what arises from God's substance is realized here. The «Big Other» must necessarily be outside and beyond this world, and that other and different cannot be related to the characteristics of beings but instead to Being as preceded by God. In scholastic logic, as is known, which continues with Aristotelian logic, the term *tertium non datur* or «the third does not exist» indicates that it is not possible for a being to exist both here and there at the same time and that the truth of Being appears only in judgement in the sense of matching opinions and things themselves such that only one statement can be confirmed, the other being untrue or false. The fundamental logic of metaphysics is either/or logic.

But from Plato to Georg W. F. Hegel, Karl Marx, Sigmund Freud, Jacques Lacan, and cybernetics, we encounter something self-evidently uncanny: *Unheimlich*. Plato established the triad or trinity of theory, practice, and production. The latter is established as *poiesis* and signifies the production of beings from Being. Based on this assumption, Christian theology holds that God is the creator of all things (*creatio ex nihilo*). *Poiesis* denotes a creation, work, and production by which man shapes his world as a work of necessity and freedom because production in the active and non-active sense, for example, poetry and sculpture, is something «innate» to him that determines his essence. Of course, production ranks third after theory and practice. Hegel's absolute is the highest rank and possesses the characteristics of totality in historical-dialectical progress and development.

Thus, philosophy denotes a condition for the possibility of creating absolute science and is above art and religion. For Hegel, the «Big Third» appears as the essence of absolute metaphysics and as the end of history, in the sense of the termination-overcoming [*Aufhebung*] of the previous two stages of historical development: nature and society, and subjective and objective spirit. Only at the third stage does the meaning of philosophy as metaphysics become severe, and history assumes the character of a theodicy of the world spirit, which knows itself as the truth of the entire process of events from beginning to end. Marx, on the other hand, gets the «Big Third» through the so-called triple pattern of the rule of capitalism from the first stage of goods through money to capital or the pure idea of the actual process of the development of world history, which even prevails in what Vanja Sutlić calls the practice of work as scientific history, and that is nothing more than communism as the highest form of the meaningfulness of history in general (Sutlić, 1974).

Human, both in Hegel and in Marx, is a free individual who exists only in synthesizing the «Big Third» and history as the conception of the idea's movement through the necessary characteristics of its appearance. Communism denotes the «immanent transcendence» of the historical absolute in which man exists. Still, communism is not the goal of history but its end in the metaphysical sense. Freud discusses the subject's consciousness stage in terms of the *Id, Ego, and Superego*. Of course, the fundamental problem of psychoanalysis is the «Big Third», which commands and oppresses, liberates and oppresses, becomes an insurmountable obstacle for the free development of a person, or has the value of faith and hope in a future society of happiness, well-being, and all-round personal development. Lacan «deconstructs» this same scheme, yielding the imaginary, the symbolic, and the real, which is traumatic because it represents a split between the first and the second and exists only as a desire to reach a sublime object. The «Big Third» within metaphysics is, therefore, the initiator of history, its goal in the sense of the synthesis of substance and subject, a moment that is found in the actuality of the «here» and «now» and not in the mythical past or the indefinite future.

To be third means to be that which transcends human existence in terms of form, but it is found in it in terms of content and determines the limits of its activity. Since humans do not have their own eternal and permanent «essence», the «Big Third» is always metaphysically thought of as Kantian regulative action in the sense of human historical perfection, self-consciousness, self-organization, and self-rule until the transition to a state where the character of the human-too-human no longer exists is complete. God, history, the absolute, work, and what goes beyond the limits of metaphysics and appears in the trans-classical logic of the technosphere, i.e., the virtual actualization of becoming, are simultaneously the transcendence and immanence of life itself, which, for Aristotle, was already synonymous with Being.

However, the difference lies in the fact that the technosphere, as the third order of cybernetics, signifies the rule of autopoiesis, which thinks and no longer operates through the stages or historical platforms of the past. Still, its development indicates the irreversible hyper-plasticity of artificial life in the eternal present *(nunc stans)*. From now on, we can only conditionally talk about the first, second, and third terms within the speculative and reflective triad because the idea of history as eschatology and soteriology, as theodicy and messianism in all imaginable versions from Karl Marx and Walter Benjamin to Emmanuel Lévinas and Jacques Derrida, is no longer valid for the technosphere. We no longer value the «third» from dialectical logic as the highest in the rank of things. No, all that lies behind us belongs to the musealized past of thought.

The paradigm has three correlative meanings for understanding the technosphere. All three are at the same level of conceptual-categorical redundancy and are almost tautological. The reason for this is that the technosphere no longer has the metaphysical meaning of Aristotle's logic as an *episteme téchne*; instead, it is an autopoietic construction of an artificial reality that does not exist in the so-called first or actual reality. When Simondon, the most crucial philosopher of cybernetics, argued that the concept of information cannot, by analogy, be identical with Aristotle's form in the senses of eidos and *morphé*, it marked a reversal in the essence of metaphysics. The reason was that, with the emergence of the thinking machine (i.e., the computer), the model or matrix of Western philosophy as an ontology changed.

«Paradigm» should therefore be freed from its Greek metaphysical meaning, even though the term is historically Greek, like almost all others we still use in philosophy and science today, albeit with altered meanings after Latin dominated Western civilization in the era of Rome and Christian scholasticism. Hence, these three meanings of the paradigm belong to an area that is no longer bounded by the relationship between philosophy, science, and art in the thinking of Being, traditionally speaking, but is open-closed in the «black box» of cybernetics, which, in Hegelian terms, presupposes the thinking of a thought.

The paradigm, therefore, must be understood as (1) the transversal thinking of the technosphere as a contingent event of the creation of a new artificial life; (2) a matrix or a cognitive-theoretical-pragmatic-production framework in which thinking is articulated throughout history as linguistic, the visual and numerical code thus becoming a universal tool for creating and retrieving reality in thoughts; and (3) a conceptual-categorical system of thinking is shown as a fundamental structure, such as idea in Plato, energy in Aristotle, the spirit in Hegel, work in Marx, event in Heidegger and Deleuze, theory of relativity in Einstein, theory of black holes in Hawking, and technological singularity in posthumanist Kurzweil (Paić, 2023).

6.4 Conclusion

The thinking of the technosphere is neither a thought «about» something nor an opinion «on» something. Instead, it is paradigmatic thinking in the form of a cybernetic circle of circles, through which artificial intelligence constructs an artificial life from a trans-classical logic that combines the fourfold information-feedback-control-communication into a «transversal order» of meaning. «Transversal» in Latin denotes a transverse direction intersecting two other directions. In a figurative sense, we can refer to a crossroads as a point that shortens the distance between two places. Wavy motion in cosmology serves as a paradigm for how thought is expressed in the three meanings above. That is why «paradigm» is a term used in philosophy and the theory of science, denoting a shift from a previous paradigm in understanding the universe's origin and essence, particularly with respect to the relationship among matter, energy, and information. For example, the difference between Ptolemy and Copernicus lies in their understanding of how a paradigm emerges and operates as a universal communication system between actors in the human world.

The same applies to the difference between Isaac Newton and Albert Einstein. However, the technosphere cannot be limited to unconditional progress in understanding the scientific picture of the world; rather, it is also—and primarily—about an open-closed order of thought figures that are not mirrors of reality, but conceptual holograms interconnected with other such tools of thought. A paradigm denotes a fundamental structure by which we think about matrices in historical and epochal terms, such as Platonism and Aristotelianism in the Renaissance, or Nietzscheanism and Heideggerianism in postmodern or twenty-first-century philosophy. It is the power of designing disorder and cuts in the historical development of thought rather than order, thereby producing a paradox. Why? The technosphere paradigm should be understood as a transversal journey through emergent events of chaos and contingency. This journey becomes a post-metaphysical wandering without a first cause or a final purpose. What else does wandering mean besides an event of an entirely different matrix or paradigm of thought in the contemporary world?

References

Bailey, K. D. (1990). *Social entropy theory*. SUNY Press.

Beck, U. (1992). *Risk society: towards a new modernity*. SAGE.

Blumenberg, H. (1985). *The legitimacy of the modern age*. The MIT Press.

Deleuze, G. (1992). Postscripts on the societies of control. *October, 59*, 3–7.

Deleuze, G. & Guattari, F. (1994). *What Is Philosophy?* Translated from the French by G. Burcell & H. Tomlinson. Columbia University Press.

Descartes, R. (1993). *Meditationes de prima philosophia*. Translated from Latin by T. Ladan. Demetra.

Foucault, M. (1994). *The order of things: archaeology of the human*. Vintage.

Heidegger, M. (1954). *"Die Frage nach der Technik" in: Vorträge und Aufsätze*. Klett-Cotta.

Heidegger, M. (1977). *Sein und Zeit* (Vol. GA.2). V. Klostermann.

Heidegger, M. (1989). *Beiträge zur Philosophie (Vom Ereignis)* (Vol. 65). V. Klostermann.

Heidegger, M. (1997). *Besinnung* (Vol. GA.66). V. Klostermann.

Heidegger, M. (2003). Die Zeit des Weltbildes. In *Holzwege*, pp. 75–113. V. Klostermann.

Heidegger, M. (2009). *Leitgedanken zur Entstehung der Metaphysik, der neuzeitlichen Wissenschaft und der modernen Technik* (Vol. 76). V. Klostermann.

Paić, Ž. (2018-2019). *The Technosphere* (Vol. 1-5). Sandorf and Mizantrop.

Paić, Ž. (2022). *Art and the Technosphere: the platforms of strings*. Cambridge Scholars Publishing.
Paić, Ž. (2023). *Superfluity of the human: reflection on the posthuman condition*. Schwabe Verlag.
Severino, E. (2013). *Capitalismo senza futuro*. BUR Rizzoli.
Simondon, G. (2017). *L'individuation à la lumière des notions de forme et d'information*. Jérôme Millon.
Sutlić, V. (1974). *The practice of work as scientific history*. Kulturni radnik.
Wittgenstein, L. (1958). *Philosophical investigations* (2nd ed.). Basil Blackwell.

CHAPTER 7

War, the Technosphere, and the Question of Evil Three Paradigms of War

7.1 Introduction

There have been three paradigms of war since the dawn of Western metaphysical history. With a minor deviation, this also applies to the rest of the world, which until the twentieth century had not yet reached what Hegel, in his speculative-dialectical philosophy, called the substance-subject of world history. The first paradigm, from Heraclitus to Hegel, is characterized by the insight into the first cause and the ultimate purpose of history as a meaningful course of events, in which the destruction of nature and humanity serves to some "higher" goal arising from the onto-theo-cosmo-anthropological organization of metaphysics. From the Greeks and Romans to the Jews and Christians, this goal is always marked in different ways. For the Greeks, from the Trojan War to Alexander's conquest of Persia and Egypt, war, as the realization of the syncretic spirit of Hellenism, represents a heroic act of freedom by which the people of philosophers and poets defend their homeland and the honour of the *polis* against the barbarian aggressors from the East. As the borders of the *polis* expanded, a new logic of the imperial order emerged, with Rome embarking on a mission to civilize the world through the brutal destruction of the Other and its complete subjugation to the *Pax Romana*. Therefore, in the ancient world, war was seen as a necessity for conquering and preserving freedom, and the colonization of the Other had the Janus face of the civilizing process (Elias, 1976).

Ž. Paić, *The Technoscientific Turn of Philosophy*,
https://doi.org/10.1007/978-3-032-21081-4_7

In other words, its bright and wild thoughts conflict: i.e., ancient colonization conceptualized both political freedom for the people and the rule of an empire in which all other states are merely limited territorial colonies. So, Heraclitus lived in Ephesus (modern Turkey), and Emperor Augustus exiled Ovid to a distant island in the Black Sea, where the Roman Empire's last stronghold on the border with the Scythians was located. The metaphysical paradigm of war is thus mediated by the primordial and initial [*arché*] and the last and final [*eschaton*].

The emergence of war in the mythical-religious sense stems from hurt human vanity [*thymos*], as in the case of Achilles' anger that Homer sang about in the *Iliad* (Sloterdijk, 2006, pp. 6–12). In this way, the human in this other world appears even more radically expressed than in the worlds of the gods on Olympus or in Hades, because the sky and the earth were created by the primordial struggles of the Titans for supremacy over the world. However, the divine-human history of warfare between nations and states could not have been possible without the singular heroes who, through their tragic sacrifices or destinies, from Achilles to Odysseus, fulfilled the mission of a meaningful history of humanity. All wars are, therefore, justified from this mythical-religious frame of reference and are necessarily both conquering and just. But, in Nietzschean terms, no point of view about war and its ultimate meaning can be absolute. Matters should be judged through perspectivism.

For the Greeks, the Trojans were the cause of the war because they kidnapped Helen from Menelaus and took her away from Greece. Furthermore, the Greeks characterized them as barbarians and as prone to despotism, so their preventive destruction and expulsion from their own country were cast as an act of an enlightening and civilizing mission. From the perspective of the Trojans, on the other hand, the Greeks represented the worst forms of inhumanity and crimes against humanity because they exterminated a small nation and its centuries-old culture through war. Therefore, the question of the meaningfulness of war as the origin and materialization of the "new" in history always marks its instrumental nature. War has never meant anything in itself [*an-sich*], because then it would be utterly meaningless, akin to the apotheosis of eternal destruction and the rule of chaos. At the same time, it would be metaphysically beyond the power of gods, nature, and people, and that is truly impossible in the way of thinking that Derrida calls «logocentrism» (Derrida, 1967).

The Judeo-Christian paradigm of war is no longer determined by this instrumentalization of the concepts of freedom and mind as a metaphysical justification for colonizing the Other. Instead, it is entirely a question of the messianic liberation of the earth for the coming of the Kingdom of God. The so-called holy war serves as a model of the eschatological vision of the end of history, which is not the goal of history, as the famous aphorism in Benjamin's early *Theologico-Political Fragment* (1921) claims (Benjamin, 1972, pp. 203–204).

For the Greeks, the gods, the protectors of heroes, were in a state of permanent war and always participated in war games. Athena, the goddess of wisdom, in the immortal statue of Phidias in the Parthenon, holds a spear in her right hand and an owl in her left, whereas Ares, the god of war, is a monstrous figure from mythology whom it is unwise to anger.

This Deleuzian "immanentism" of life as an eternal conflict in the will to power cannot be the essence of the matter for Judeo-Christian soteriology. The change begins with the notion of the divine foundation of the world, in which messianic redemption becomes the ultimate goal, so holy war cannot be the ultimate objective of history. Still, it serves only as a means for, for example, Christ's soldiers or knights to achieve victory over "unbelievers" and proclaim the triumph of God's justice and love. Therefore, God does not interfere in the events of war because his kingdom rests outside this sphere. Every war within the framework of this "great narrative" of Judeo-Christian metaphysics is holy, and its meaning derives from the justification of what Hegel asserted at the end of his *Philosophy of History* when he said that the end of history denotes the realization of the theodicy of the world spirit (Hegel, 1970, p. 540).

Metaphysics and the philosophical conception of war within this model of Western thought necessarily come to an end. We could even add something apparently "derogatory". In Hegel's eschatology of world history, Greco-Roman colonialism and imperialism (with the idea of war as a natural state), as well as Jewish-Christian messianic metaphysics (with the most significant thought reach of the cosmopolitan world order in Kant's *Perpetual Peace*), were dialectically terminated and overcome (Immanuel, 1903, p. 142).

Following the demise of Western metaphysics' core concepts and ideas, such as Hegel's philosophy of history, state, and law, war emerges from the sphere of natural-political normality between peoples in the world, becoming not only an insult and contempt for the mind that rules reality but also nonsense. Hegel concluded his reflections on world history by situating the liberal notion of English capitalism as a sign of a new era in which economics would assume the task that politics had held for centuries.

Another paradigm of war results from the destruction and deconstruction of Western metaphysics. Is it not astonishing that within the framework of this first paradigm, with the rule of *logos* and the essence of man as *animal rationale*, it is impossible to determine the nature of evil in history in any other way than as the negation of good? And since all good in the ethical-political sense is the result of the rule of the mind, then, as Arendt and Lévinas well understood, the outrageous "banality of the evil" of the Nazi system for the extermination of Jews in concentration camps cannot be explained by the essence of Kant's regulative action mind in society and the state. Instead, what evil and modern total war belongs to must be introduced into this irreducible realm beyond the opposition of mind and no-mind. As such, it leads to the ultimate consequences of horror and absolute inhumanity. What, for example, the practice of the Holocaust introduced into world history is called "radical" or "diabolical evil" (Arendt, 1973; Lévinas, 1988, pp. 156–167). In this second paradigm, therefore, the concept of war does not consider the necessity of the forces of destruction in nature and man, aided by methods and technology, for history to be meaningful in its eschatological "secret mission".

Moreover, the most significant thought representatives of this circle of destruction and deconstruction of metaphysics, such as Heidegger, Schmitt, Benjamin, Arendt, Lévinas, Derrida, and Patočka, tried to open the question of the possibility of a different paradigm of war and, at the same time, the concept of evil, without which it is not possible to continue talking about the philosophical understanding of war without problems. This, of course, does not mean that evil in the theological-political sense appears as a non-reversible Kantian modal category, that is, as a condition for the possibility of war. Such a pseudo-transcendental relic of metaphysics is no longer relevant for any modern political theology. Evil is not "outside" this world and does not "enter" this world through what Malle-branche calls occasional causes, but rather in the sense of the negation of God's presence. In contrast, the relationship between evil and war does not involve an absolute causal connection. Schelling has already explicitly emphasized this in his reflections on human freedom. For example, in his definition of the difference between the essence of human freedom and "human freedom" itself, the question of «freedom of will, Good and Evil, personality, etc.," appears (Schelling, 2006, p. 15).

To be able to arrive at a proper establishment of concepts such as evil and war outside the perspective of Western metaphysics, we must apply both evil and war in the thinking of what has happened throughout the entire history of humanity, culminating in the concepts of total war and war carried out with the help of the technosphere as the realized essence of the cybernetic foursome (information-feedback-control-communication) in the explosion of energy as mass and the implosion of information as chaos and entropy, to the ultimate possibilities of reduction to "liberate" any negative messianism and apocalypticism. As Clausewitz wrote, war is not the continuation of politics by other means, which Heidegger "undermined" in his thinking, starting from the fact that war is the occurrence of an "accident in the abyss of life" (Heidegger, 2014, p. 152).

Nothing in the entire legacy of metaphysics helps us think about what I am trying to bring to light as a problem of the relationship among war, the technosphere, and evil (Paić, 2018, pp. 40–54). The autonomous consideration of the possibility of what I call here the third paradigm is unworthy. In the first paradigm, the technical notion of metaphysics and its understanding are at work, which determines philosophy by starting from traditional thinking in logic, causality, and teleology. In this framework, the meaning of Being, the revelation of God, the understanding of the world, and the position of humans are correlated but arranged in a distinct vertical hierarchy. War always denotes this or that war, just or unjust, heroic or cowardly, exterminating and genocidal, and since it is always in the function of something else, namely the creation of a nation-state, the conquest of new territory, and the liberation of humanity from the natural shackles of tribes and clans by necessity, it seems self-evident that, like the concept of technique, it should be reducible only to the pure instrumentality of transcendence and messianism. All wars of this metaphysical rank, fought from the Trojan War to the First World War, are determined by the geopolitics of primitive (re)territorialization, to use the terms of Deleuze and Guattari (Deleuze & Guattari, 1980).

War, conceived from that second paradigm, represents a total technological device for the destruction of land, resources, and people in the name of a metapolitically understood nation-race-humanity. For the first time, therefore, war appears as a means-end of the inhuman in the very "essence" of realized metaphysics, so it is not surprising why the

First World War, in the jargon of the German conservative revolution, in the writings of Spengler and especially Jünger, received the attributes of the total mobilization of destructive technology, such as the long-range artillery that tore human flesh in the trenches of Verdun. Technology transcends the boundaries of nature and the essence of land and human interaction, as seen in handicraft. The mechanical technique still evokes what Nietzsche called human-to-human interaction; wars also convey a romantic illusion of chivalry and hand-to-hand combat, much like the conflicts of the tragic heroes in Homer's epics, especially Achilles and Hector. Therefore, war as a mystical, existential experience is celebrated by none other than Jünger, in the footsteps of Nietzsche and avant-garde, futuristic, and expressionist aesthetic visions of the world (Jünger, 1926).

Technology becomes what is freed from human presence. Still, without its surveillance and driving power as a will to kill the Other in the name of the nation-race-humanity, there is no possibility of self-directed action at a distance. One can read about this in various texts by philosophers of life, such as the anthropologist Scheler or the proponent of political decisionism Schmitt (Scheler, 2016; Schmitt, 1997). The key point of differentiation in opinions about war within this second paradigm is that war is viewed as the total mobilization of capital, technology, and social militarism, rather than the romantic conquest of the Other's territory, even in the nineteenth century. Regardless of the absolute policy of genocide against African peoples in the Congo ordered by King Leopold I, this new technology of destruction, created by the assemblage of mechanics and engineering within the flourishing field of thermodynamics, perfectly coincided with the pursuit of modern imperialism, which meant the end of the fragmentation of the European concept of the nation-state.

In this sense, the technological dispositive of the destruction of the earth and the performative event of the inhuman in the very "essence" of evil continued in the Second World War, which, from the point of view of the method of action, was a total war of the elemental triad: namely, the sky (aviation), the earth (armoured vehicles like tanks and human forces, artillery and missile strikes), and the sea (ships and submarines). Most importantly, however, the concept of total war is not so much in a war's expansion in the quantitative field of increasing the number, strength, ratio, and scale of the destruction of object *X* but in the qualitative power of conquering space-time to establish its platform of progress in the production of inhumane. The essence of total war derives from five post-metaphysical "attributes" of contemporary nihilism. It thus connects all three paradigms of understanding war (the technical one, the technological one, and the one I call the technosphere) as a combination of artificial intelligence and artificial intuition. In 1938, Heidegger interpreted this set of relationships from a philosophical perspective, without which it would not be possible to think of the posthuman condition of total war. These are:

(1) the *dynamic* that refers to the development of strength;
(2) the *totality* that affects the principle of power so that nothing outside that sphere of action can maintain its innocence and still be considered "real";
(3) the *imperial* one that results from the commanding character of power and suppresses any possibility of exception in one's environment.
(4) the *rational* one, in which the computational character of thinking is recognized in the closed framework of execution forces, and
(5) the *planetary* one, which shows that power can no longer be "total" but is bound to the limits of the inhabited globe, such as the atmosphere and stratosphere, which means that the image of power conquers through the "penetration" of the planet as a whole and thus neutralizes possible planetary opponents. (Heidegger, 1997, p. 18)

The third paradigm of war represents the end of war in the (post)metaphysical sense. It is no longer the methods and technology of destroying the earth in the name of the messianic-political goals of justice of the eternal community or the perpetual empire of the nation-race-humanity, as were the political goals of totalitarian fascism, Nazism, and Stalinist communism in the twentieth century. Instead of this metapolitical advocacy of the rule of the subject-substance of "its" idea, which spreads imperially and colonially around the world with the aspiration of turning it into an absolute zone of geopolitics, as can be seen in the twenty-first century in how Putin's brutal invasion of Ukraine threatens the logic of the global-planetary politics of reorganization, we have found ourselves for the first time in a pure situation that derives the relationship between evil and war from the all-encompassing "positivity" of the will for power as the destruction of the human species or humanity.

However, this new Armageddon does not come with any external cosmic catastrophe, nor with the so-called natural extinction of this species, nor by mutual human-to-human destruction, potentially helped by nuclear war or wars between the existing empires (USA, Russia, and China) and their colonial satellite states. Instead, the third paradigm of war is determined by the total and absolute rule of the technosphere, which enables the emergence of artificial life from the techno-genetic construction of artificial intelligence. All this began with the SF idea and imagination of a futuristic scenario in the 1960s with Kubrick's film *2001: A Space Odyssey*, in which the superintelligent computer HAL 9000 kills the ship's crew until the astronaut Bowman manages to deactivate it at the last moment and thus saves hypothetical humanity in its post-and-transhumanist transition to a higher stage of cognitive evolution (Paić, 2023, Sorgner, 2023).

In the metaphysical sense, war is the realization of the idea of ruling over the Other, as in the Greek paradigm of the political and politics, which, despite formal freedom, equality, and justice for Greeks and citizens of the *polis* as foreigners, does not include enslaved people and women in the totality of public recognition. In addition, war should be understood in terms of the correlation between psychopolitics, both as a singular and a collective phenomenon, since it is the sphere of *thymos* that demands *ressentiment*. Any explanation of the origin of war that is explicitly economic or materialistic, as in vulgar Marxism, with rare exceptions such as Foucault and Deleuze and Guattari in the area of post-structuralism, fails to see that the strategy of violence and aggression cannot be just an accumulation of political factors that correspond to the structure of contemporary capitalism based on information and communication networks and their endless process of the emergence and contingency of capital (Allez & Lazzarato, 2016).

The same applies to the irreducibility of the third paradigm of war, which is not a war against all, as suggested by chaos and entropy in today's corporate structure of oligarchic world rule. There have been periods since the Second World War that have resembled periods of peace and a striving for non-war solutions to economic and political issues. However, these tended to be periods of so-called "stand-by" or periods between two wars, aiming, for example, for national independence and state sovereignty or, on the other hand, for the imperialist conquest of other people's territory. These may be unfinished ethnic conflicts, such as the ongoing violent conflict between Israelis and Palestinians or ethnic wars in Africa, or frozen conflicts of a warlike nature, such as between India and Pakistan in the Himalayas. However, none of this is decisive for the third paradigm of war, which, by the minimal definition, is not an aggressive conflict until the Other is destroyed and its territory is controlled.

7.2 A Cybernetic System of Transition

The technosphere, as a cybernetic system for transmitting, storing, controlling, and creating information, serves as the origin of the so-called "third machine", which unites technique and technology through algorithmic processes of calculation, planning, and event construction. The reality, therefore, becomes a techno-genetic or digital construction that Deleuze calls «virtual actualization» in his book *Difference and Repetition* (Deleuze, 2011).

The third machine signifies that it is no longer merely an extension of a human's arms or body in its technical form. Instead, it is mediated by the brain, which visualizes concepts as images, without a model or an original in nature or in the external world of culture. These are images of events, such as film images, that do not imitate or represent anything with respect to Being, beings, or the essence of humans in traditional metaphysics. These images are not even a visual language, as semiotics has sought to understand since Barthes, with its belief in language as the fundamental "signifier". In the cybernetic model of ontology, instead of the necessity and permanence of Being, whether considered phenomenologically or psychoanalytically as noema or the unconscious, we encounter the triad of contingency, emergence, and deterministic chaos. That is why vision, as a visualization of events, precedes the program as a system of computer intelligence. This antecedence, by analogy, belongs to the apriorism of speculative German philosophy, primarily Kant and his transcendental perception of space and time. However, since computer intelligence is not a mimesis of the speculative mind, human intuition cannot serve as a generative model for possible artificial intuition. Many software engineers responsible for the optimal performance of artificial intelligence's work and actions have been experimenting with this concept since Minsky (Minsky, 2007).

Since the technosphere at the current level of cognitive development is not yet a fully autonomous thinking machine, everything is moving towards the fact that it will very soon be the realization of the SF vision of Kubrick's HAL 9000, which means that, in addition to the infinite speed of calculation, planning, and construction, it will also have developed emotional intelligence and perhaps artificial intuition. As such, the question of whether an ethical understanding of the world in terms of consciousness, conscience, responsibility, and the distinction between good and evil is possible is no longer entirely futuristic and unreal (Bostrom, 2013, p. 22).

The problem with the technosphere's thinking on war, however, is that it cannot be considered an ethical issue. Here, it is no longer a question of a potential struggle or battle for the territory of a nation-state or even for supremacy over the global-planetary space. Instead, "war" becomes merely cognitive, a matter of ruling over the human mind and body as a free territory of existential performance.

Furthermore, suppose the means and metaphors of warfare in the age of technique were the sword and shield, and, in the age of technology, the cannon and the bomb. In that case, it is apparent that, in the age of the technosphere, the instruments of destruction of the Other, Earth, and humanity are nothing more than gamma rays and hypersonic missiles. There are, therefore, only two possibilities within this third paradigm of war after the end of metaphysics and post-metaphysics. The first is already at work and refers to the human use and abuse of thinking machines as intelligent or autonomous, yet destructive, weapons. The second is what can be expected: the technosphere adopts an "enactive" approach to evil and destroys the apparently advanced, technologically developed human civilization. Of course, there is a third hypothetical possibility: that, like Heidegger's last god, AI will save us from ourselves, that is, from our deadly *thymos* and destructive ressentiment, becoming a higher form of that Kubrickian "star child" on the other side of the tragic (post-) biological history of humanity.

7.3 Thinking as Techno-nihilism: Penrose vs. Severino

Since the beginning of the new era, with what Sloterdijk calls «Copernican mobilization», we have encountered an attempt by philosophical thought to construct the world from the logic of technical production rather than to observe Being as the subjectivity of an object (Sloterdijk, 1987).

For Severino, the origin of this modern-day technical nihilism lies in Plato's concept of *poiesis*. Among the numerous works on artificial intelligence and its implications for life in today's information society, it is worth highlighting the book of conversations between Severino and Nobel laureate Roger Penrose, one of the most renowned contemporary mathematicians: *Artificial Intelligence vs. Natural Intelligence* (Penrose & Severino, 2022). What is interesting here is that the conversation necessarily recalls a long-standing problem in the history of Western metaphysics: how to establish a genuine dialogue about the essence of the things at issue in the dispute between science and philosophy. Did this dialogue not begin in the pre-Platonic era with the Greeks—for example, between Thales and Parmenides—and continue in the new era with the conversation between Newton and Kant, culminating, of course, in the twentieth century with the dialogue between Einstein and Bergson?

The question, however, is what all these meetings between scientists and thinkers testify to if not the fundamental issues of ontology, such as the question of whether the world is perpetual or was created by the Big Bang, whether God exists or whether his existence was unnecessary for the creation of the world, and whether the capacity to think is only human or whether it also belongs to the non-human in some way still unknown to us. When Moravec published his visionary book *Mind Children: The Future of Robot and Human Intelligence* in 1988, he predicted that by around 2030, man would be essentially "overcome" by the decisive action of intelligent machines. This prediction still carried a streak of utopian rapture, reflecting the technical constellation of modernity (Moravec, 1988). In other words, Moravec predicted that by the mentioned year, there would be an equivalence between the natural mind and artificial intelligence. In *The Superfluity of the Human: Reflections on the Posthuman Condition*, I addressed the relationship between two modes of thinking and their complex configuration of concepts, beginning with an understanding of the *technosphere* (Paić, 2023).

In short, between thinking as rationality-planning-construction and thinking as a telling event of the meaning of Being, there is a correlative relationship and, at the same time, an incompatibility in the essential settings of what Heidegger calls the «task of thought". I tried to develop the same problem extensively in such a way that I arrived at the proposition that the future of philosophy lies in contrast to the absolute rule of the technoscientific system of the visualization and mathematization of thought to defend what enables the essence of thinking as such, which means the freedom to decide on one's path beyond the triumph of scientific positivism and the return to the abyss of primordial metaphysics. It seems that this conversation between Penrose and Severino truly represents, as Borges would say in one of his literary parables, a mirror of eternity and a trace of the metaphysical passion for truth.

But what exactly does "artificial intelligence" mean in contrast to the natural mind? Penrose, Emeritus Rouse Ball Professor of Mathematics at Oxford University, argued that artificial intelligence is a term (albeit imprecise) to describe machines that think by reproducing the activities of human intelligence. He claims, however, that all these outstanding achievements in the development of artificial intelligence, such as acceleration in data processing, the evolution of machine learning, big data, and algorithms, are still not enough to explain the true triumph of artificial intelligence over natural or human intelligence. Why? Penrose believes that the logical structure of this non-human thinking needs to be presented and therefore directs attention to Gödel's theorem on the impossibility of an "intelligent machine". However, the key problem for him is not logical but physical. In his books, he consistently emphasized that to comprehend the essence of the universe and its origin, it is essential to consider the creative dimension of the purposeful correlation between matter and energy. However, the problem of the origin of human consciousness and self-awareness remains unsolved. According to Penrose, our human mind cannot be algorithmically programmed in performance processes. Not only do we still know so little about it, but we also cannot interpret it accurately within the limited cognitive frameworks available to us. In other words, we lack a conceptual framework to address the aporias of the relationship between the natural and the artificial.

When everything is reduced to the right measure, it seems that Penrose's scepticism towards what Moravec calls «human equivalence» denotes the result of something that German speculative philosophy from Fichte to Schelling has already considered: "artificial intelligence" cannot be an adequate term for the very act of developing a thinking machine until it produces what is called self-consciousness. However, the question is whether the "self-consciousness of an intelligent machine" can only be a copy of human self-consciousness or whether we must find a different solution for what Gotthard calls «trans-classical logic» that goes beyond the principles of identity and sufficient reason (Günther, 1976, pp. 249–328). Unexpectedly, in his interpretation of this problem, Penrose argues that AI cannot be genuine intelligence because, in his view, only human intelligence is genuine.

Here, I would like to add the following. To understand the relationship between human thinking and what we call "artificial intelligence", something more than classical ontological categories like "reality" and "non-reality" is needed. As is well known, Kant discusses modal categories in which the substance and essence of a thing appear in space and time. These categories are possibility, facticity, reality, and necessity. When we discuss artificial intelligence, it is evident that, despite Gödel's theorem, a thinking machine will increasingly be supplied with information beyond human memory capacity as it develops. Plato still considered memory one of the fundamental features of human thinking because it was only through this mode of thinking that humans developed self-consciousness in space and time. The possibility of self-development in artificial intelligence rests on two assumptions: the self-development of the rational structure or logic of this thinking, and the self-development of what is called artificial intuition, which produces information by functioning according to the model of a visualized concept or image. That picture is anything but realistic, so Penrose is only partly right. Artificial intelligence is not "real" but "hyper-real".

It is precisely for this reason that the modality category in the self-development of artificial intelligence, on the path to technological self-awareness, denotes the connection between possibilities and reality as a new contingent necessity. Deleuze calls this «virtual actualization» (Deleuze, 2011). Although Penrose believes that the concept of artificial intelligence should be freed from the matrix of social implications and consequences for human communication, it falls into the abyss of reductionism when attempting to solve these issues. There remains a problem that, even at this level of artificial intelligence use, is already irremediable. The openness of space to the creative dimension of human thinking and innovation does not exclude the possibility that artificial intelligence can think and act, creating solutions that resemble human thought and existence.

In his conversation with Penrose, Severino consistently defended his views on the original structure of thought, which posits the openness and persistence of Being in its eternity, drawing on Parmenides and pre-Socratic ontology. For him, the decisive problem in the discussion of the relationship between the natural mind and artificial intelligence boils down to the fact that the conceptual framework of both is present in what Plato calls *poiesis*, i.e., production. Since the West denotes the absolutization of production precisely as the ruling and only authoritative way of thinking, *poiesis* is the cause that enables the creation of something from non-existence into existence. We can see from this how Severino tries to show that the question of thinking as a dispute between the original and the non-original lies at the very foundation of Western metaphysics. For this reason, the question of the possibility, reality, and necessity of artificial intelligence simultaneously becomes an ontological question about the world and humans. This is premised on the assumption that, for the first time in history, intelligence can become entirely non-human and therefore be engineered as a manipulative framework for subsequent action.

In contrast to Penrose, who defined real or natural intelligence in terms of conscience as a quality of consciousness that assumes an ethical dimension, distinguishing between good and evil, Severino concludes that consciousness encompasses areas that artificial intelligence cannot reach. Thus, it seems that the production of consciousness is the fundamental issue today. Hence, both the mathematician and the philosopher are aware that this conversation is nothing more than an exchange of views on the hope that the self-development of artificial intelligence cannot replace the irreducibility of human thinking and that, in the future, the relationship between the synthetic and the natural will still be in the coherence of both, not in the total power of artificial intelligence over man.

Everything that emerges from this book is not far from Heidegger's notion of releasement [*Gelassenheit*] against the technical system in the sense of preserving the property of thinking as expressing the meaning of Being (Heidegger, 1985). However, it seems to me that this is just an attempt to calm people before the unstoppable process of technologizing thoughts and visualizing events. The problem is no longer merely in pointing out that production, in the sense of the total output of consciousness, marks the path of Western nihilism; the term technosphere is intended to denote concepts that extend beyond scientific and philosophical ways of thinking. The technosphere cannot be what Penrose and Severino are discussing when they sceptically attempt to protect the irreducible territory of human consciousness. They do this with the help of ethics and ontology, synthesizing all five senses in human history, with language as the first and last guarantee. When the logic of an infinite number in computation is entirely ruled by the concept of image, producing information as visualized concepts, the problem lies in what we attribute to Aristotle as being human, namely, opinions in terms of *animal rationale*. From the essence of the technosphere as the self-awareness of autopoietic machines, it necessarily follows that thinking synthesizes the non-human into the human, i.e., the rational into the intuitive.

Therefore, it is no coincidence that, at the beginning of the twenty-first century, intensive research into artificial intuition in robotics and engineering extends beyond the limits of human cognition, which we usually attribute to art rather than science. In the conversation between Penrose and Severino, the "third man"—the synthesis of artist and visionary—was missing: the "great third", which is neither God nor technology, but that which can think differently and creatively produce something new without it being a mere imitation of human creativity. Is this *tertium datur* still only a possibility and not a reality and a necessity, or perhaps, has the virtual actualization of the non-human as a creative designer of worlds already been realized?

Why does the conversation between mathematicians and philosophers seem so crucial for understanding the relationship between the technosphere and war? Because it shows that thinking as calculation-planning-construction versus thinking as mythopoetic telling signifies the primacy of the pragmatics of knowledge about the possibility of mastering the world as the conquest and subjugation of the earth and the universe with the fundamental purposes of what constitutes the essence of modernity, which is the rule of the subject-substance through progress and the development of the technoscientific objectification of Being. Therefore, it seems evident that, with the advent of a new era and worldwide scientific and technological advancements, a hegemonic mode of thinking has emerged that tends to assume military features through the quantification and visualization of Being.

The military cannot be limited to the realization of war plans and programs for the complete conquest of the earth and space, driven by the power of a singular or collective subjectivity. It involves the colonization of the land and the rationalization of the world, beginning with the tendency to extend the concept of total mobilization, which Jünger discusses in his analysis of the enslavement of Being and the nihilism of technology, to the real conditions of governance beyond all limits. However, the boundary between thinking as the techno-militarism of the world, which Nietzsche calls «logic, mass and brutality", and thinking as a discursive conception of Being cannot be only a difference in the sense of a plea for war and a plea for peace (Nietzsche, 1980, p. 473).

Instead, in these two modes of human thought, we confront the final boundary in the very "being" of Human, as both *animal rationale* and *homo kybernetes*. While the anthropological reduction of the ontology has a scientific-technological character for the former, for the latter, as the "last man" in the posthuman condition, the difference between what Heidegger calls calculating thinking [*Rechnen*] and what he calls telling thinking [*Dichten*] cannot be decisive. Why? For the simple reason that for the contingent-emergent action of "man" in the environment of the technological dispositive of "artificial intelligence", any insistence on defending the irreducibility of mythopoetic narration and the humanistic set of "values" represents "reactionary" nostalgia for the old days of metaphysics. This obscure dualism between the Platonic rule of *téchne* and *poiesis* is no longer relevant in the technosphere, as it is the essence of a cybernetic system for managing a technically shaped world. As an event of the third paradigm, war represents the "natural state" of self-regulation in the relationship between the techno-military system and its environment, in which local and regional wars occur at all levels of performativity. In this respect, Jünger left the following prescient thoughts, irrefutable proof of the complete bankruptcy of all messianic-romantic illusions about the meaning and purpose of wars in the post-historical era:

> The perfection of the technical means of power consists of the ultimate state of expediency and the possibility of destruction. [...] Already in the last war, there were zones of destruction whose scenes can only be described if we compare them to natural disasters. [...] It is a romantic thought that social contracts can restrain the rampages of these energies and their application in life-and-death struggles. The premise is that man is simultaneously good and evil. In every calculation that needs to pass the exam, it is necessary to include the fact that a person is capable of everything. Moral rules do not determine reality; laws determine it. (Jünger, 1981, pp. 145-146)

From all that has been said, something uncanny emerges. First, what lies behind the third paradigm of war in the term "technosphere" can, for the first time, become a realized absolute: a triad of metaphysics, cybernetics, and transhumanism. The technological singularity is therefore not a messianic event in the cosmic-human evolution of consciousness, which, by merging with God, will lead to bliss on the other side of history. Instead of the traditionally understood soteriology, which remained the philosophical view of modern physicists, mathematicians, and monadologists such as Newton and Leibniz, there are growing concerns about the linear development model of emergent artificial self-awareness.

The reason is apparent. The production of autonomous weapon systems or "killer robots" that can find military targets and neutralize them with smart bombs without human intervention has already gained momentum in recent years. Drones are potent weapons in military conflicts. In addition to replacing aviation, their widespread use is leading to significant changes in warfare, affecting classic means of destruction, such as tanks and multi-barrel rocket launchers. No one can stop this anymore, especially not by making ethical appeals. This includes the one signed by many leading technological experts, roboticists, philosophers, and linguists, who claim to intend to prevent the so-called third revolution in warfare after the invention of gunpowder and nuclear weapons.

It cannot be stopped because this is neither the first nor the second paradigm of warfare, nor, therefore, is it the sword and shield, nor is it the cannon and the bomb, which were created as a development of technique and technology as a means of conquering the Other in the sense of (re)territorialization. This "here" and "now" are not spatially and temporally fixed and permanent in the opinion of some great inventor or designer. This arises from the technosphere, a unique human process that merges with the non-human, as in the emergence of cybernetics, with its key concept of information, which entails transcending the concepts of matter, such as mass and energy (Wiener, 2019).

For the performative enactive mind of the third or cybernetic machine, the world is no longer post-temporally "in" the world. Instead, in the middle is a techno-genetic construction in which Being, beings, and the essence of man are no longer ontologies. After all, the authoritative concepts of the new operational opinions are based on the trans-classical logic of *tertium datur*. That third—the third machine, the third hand, the "great third", the third paradigm of war—is nothing but the end of the distinction between metaphysics and cybernetics. Instead, thinking within the sphere of action of the technosphere extends beyond the earth and the territorial rootedness of things and objects, as well as the autonomous object itself, object *X*, which "thinks" by quantifying and visualizing what lies on the horizon of the immeasurable and invisible, such as technoscience with the human eye.

Artificial intelligence therefore becomes only the algorithmic potentiality of the construction of events that do not exist in reality but are virtually actual and thus hyperreal as far as the strategic planning of the future war is concerned, for now still in the heads of people and their metaphysical fictions and obsessions for the will to power against the existence of the Other as the enemy of my subjectivity and substantiality, and declare what is "new" on the *event horizon* as a legitimate war objective that should be neutralized or destroyed. We can now see that what until the 1990s had the feature of "futuristic" or SF, like the dystopian films *Blade Runner, Alien* and *Terminator*, has since become a hyperreal event of the combination of information, genes, and robotics in the production of a third machine or technosphere that "thinks" and acts autonomously but in the sense of *Rechnen,* not *Dichten.* In the future, this tendency may even evolve into a techno-mythic-poetic realm, which presupposes unimagined possibilities for aestheticizing and designing digital platforms of cybernetic synaesthesia.

7.4 Wars Without Worlds?

Any new "philosophy of war" no longer has its raison d'être for the simple reason that a war waged with autonomous weapons for the destruction of the Other, not necessarily human, is no longer a war-in-the-world that has, phenomenologically speaking, its *noema* or subjectivity of the object. The concept of the world in the metaphysical history of the West, until Heidegger, had that essential ontological difference between Being and beings, so it was self-evident to distinguish between the world as *cosmos* and *mundus* in the Greek and Christian sense, and the world as the horizon of designing the event of Being and the essence of the human. The worldliness of the world has always been a "thing" of thought within the metaphysical triad of Being-God-Human. The fourth, that is, worldly in the concept of the world, belongs to an opinion that does not reduce the absolute integrity of the world to something external to it, such as science or technique-technology, which Heidegger destroyed in *Being and Time* by asserting that the meaning of Being is missed in the modern reductiveness of biology, psychology, and anthropology (Heidegger, 2018).

"Life", "soul", and "Human", therefore, do not express the subject-centricity of thinking but are self-showing and self-expressing phenomena of the meaning of Being in the mode of an existential project of freedom as a "being there" [*Dasein*]. When the world in the age of the technosphere loses its cosmocentricity and anthropocentricity, that is, what in the term "Anthropocene" signifies its carbon structure of the earth and the end of the industrial conquest and destruction of the surrounding world [*Umwelt*], the logical consequence is that from the horizon of meaning, the world becomes astrophysically reducible to the possibility of its disintegration in the projection of the *event horizon,* which, as Hawking knew, marks the end of the possibility of all visibility and the cognitive-theoretical validity of physical categories.

When we keep this in mind, it becomes evident that the technosphere transcends the boundaries of human and the earth, and that its space-time is characterized by contingency, emergency, and deterministic chaos. The technosphere lies beyond any metaphysical picture of the world. Even Heidegger's brilliant analysis of the emergence of the scientific-technical age with the term «the age of the world picture» [*Die Zeit des Weltbildes*] is no longer an appropriate framework of thought for what is happening "here" and "now" (Heidegger, 1977, pp. 87–88).

Instead of world history, we are immersed, and no longer thrown out, into the global-planetary nihilism of the third paradigm of war beyond the war for (re)territorialization because total war in the hyperspace of Schmitt's *Großraum*, which I call *post-imperial sovereignty*, encompasses everyone and no one, with a tendency to become an interplanetary war without worlds, that is, in the singular plural, it is about permanent wars that will potentially be waged between humans, *homo kybernetes* and extraterrestrial creatures in the coming future (Paić, 2022, pp. 309–340).

This syncretism of subjects/actors shows that Jünger, Schmitt and Heidegger, not the heirs of Kant and the Enlightenment, are the thinkers who were quite clearly "down to earth" in thinking that the human is neither an angel nor devil but a being that, in its possibilities, encompasses precisely the demonic ability to realize the goals of radical or diabolical evil and, just as Schelling first thought in his discussion about the essence of freedom, becomes the fundamental issue of any further discussion about the meaning of life in the age of artificial intelligence. Is it not amazing that as soon as *ChatGPT* was released by OpenAI in 2022, all previous "great narratives" of metaphysics, from phenomenology, psychoanalysis, Marxism, the philosophy of life, hermeneutics, and philosophical cosmology to cybernetic anthropology and various theories of war from the perspective of the ethical-political renewals of Kant and cosmopolitanism, lost any credibility? All the new so-called successful positive science philosophies are rooted in neurocognitive and neurophilosophy, so this referential framework of the brain as calculation, planning, and construction becomes a reductive path in understanding war in the age of the technosphere. However, let us not celebrate what should not be celebrated, because the brain, as a vision and as a program, only reveals the ambiguous wisdom of Heidegger's statement from *What Is Called Thinking?*

> *Das Bedenklichste in unserer bedenklichen Zeit ist, dass wir noch nicht denken.* (Heidegger, 2002, p. 5)

To think of war in the age of the technosphere means to think radically differently from the ruling current of this techno-nihilistic fall into the abyss of "new positivism". War denotes the *polemos* of all things; hence, its destructive history of chaos and entropy paradoxically becomes our creative future, if what remains presupposes the openness of the question of evil rather than the closure of thought in the ever-new fiction of perpetual peace and its fallen angels of infinity. To think of what stands outside this event horizon means to think of what is impossible and inexpressible and, as such, essentially undemonstrable.

Note [1] The event horizon denotes the boundary that defines the region of space around a black hole from which nothing (not even light) can escape. In other words, the escape velocity of an object within the event horizon exceeds the speed of light. It is so-called because it is impossible to observe any event that takes place within it—it is a horizon beyond which we can no longer see anything. The event horizon, therefore, effectively hides the singularity at the centre of the black hole, a boon for astrophysicists uneasy about the breakdown of the laws of physics at such a point. The cosmic censorship hypothesis states that the event horizon hides all singularities and that the absence of physical laws is not visible. In theory, any mass can be compressed enough to form a black hole. The only condition is that its physical size must be smaller than the radius of the event horizon. For example, our Sun would become a black hole if its mass were contained within a sphere with a diameter of about 2.5 km.

The event horizon of a rotating black hole splits into an outer horizon and an inner horizon. The outer event horizon of such an object acts as a point of no return, just like the event horizon of a non-rotating black hole. The inner event horizon of a rotating black hole, also known as the Cauchy horizon, is stranger. After that threshold, cause no longer necessarily precedes effect, the past no longer necessarily determines the future, and time travel may be possible. (In a non-rotating black hole, also known as a Schwarzschild black hole, the inner and outer horizons coincide.) A rotating black hole also forces the fabric of space-time around it to rotate with it, a phenomenon known as frame dragging or the Lense-Thirring effect. Frame drag can also be seen around other massive bodies, including Earth.

In summary, the concept of the event horizon originates in Einstein's theory of relativity. Everything that happens beyond that horizon only shows that the known physical laws no longer apply, so even the model of causality cannot explain the interaction of what is beyond these horizons with what is inside. We would say that contemporary science in this respect is identical to what Deleuze claims when he introduced into the philosophy of becoming [*devenir*] what Duns Scotus, Spinoza, and Nietzsche called "immanence". The worldliness of events marks the boundary between the empirically determined world of phenomena and what cannot be known, such as the Kantian thing-in-itself [*Ding-an-sich*] or object X.

However, what if, from the point of view of future physics as a metaphysics that synthesizes experience within itself, astrophysics and cosmology, and philosophical research into the possibility of the emergence of what lies behind our knowledge, the event horizon is a temporary cognitive-theoretical boundary, not something fixed and unchanging? What, then, if events can happen in such a way that Whitehead's hypothesis that the evolution of cosmic epochs requires the creation of new conceptual frameworks for understanding the contingent necessity of a universe for which God is not the first cause and the last purpose, but possibly the creative-metaphysical initiator of the process into infinity without the last insight into its newness of eternal becoming? (Merali, 2014).

References

Allez, E., & Lazzarato, M. (2016). *Wars and capital*. Semiotext(e).

Arendt, H. (1973). *The origins of totalitarianism*. Brace Jovanovich.

Benjamin, W. (1972). *Theologisch-politisches fragment*. In H. Schweppenhäuser & R. Tiedemann (Eds.), *Gesammelte schriften* (Vol. II-1). Suhrkamp.

Bostrom, N. (2013). *Superintelligence: paths, dangers, strategies*. Oxford University Press.

Deleuze, G. (2011). *Différence et répétition*. PUF.

Deleuze, G., & Guattari, F. (1980). *Mille plateaux: capitalisme et schizophrénie I*. Minuit.

Derrida, J. (1967). *De la grammatologie*. Minuit.

Elias, N. (1976). *Über den prozeß der zivilisation: soziogenetische und psychogenetische untersuchungen*. Suhrkamp.
Günther, G. (1976). *Beiträge zur Grundlegung einer operationsfähigen Dialektik*. Felix Meiner Verlag.
Hegel, G. W. F. (1970). *Vorlesungen über die Philosophie der Geschichte* (Vol. 12). Suhrkamp.
Heidegger, M. (1977). *Holzwege*. V. Klostermann..
Heidegger, M. (1985). *Gelassenheit*. G. Neske.
Heidegger, M. (1997). *Besinnung, GA* (Vol. 66). Klostermann.
Heidegger, M. (2002). *Was heisst Denken?* (Vol. 8). V. Klostermann.
Heidegger, M. (2014). *Überlegungen XII-XV (Schwarze Hefte 1939–1941)* (Vol. 96). V. Klostermann.
Heidegger, M. (2018). *Sein und Zeit* (Vol. 2). V. Klostermann.
Immanuel, K. (1903). *Perpetual peace: a philosophical essay*. George Allan & Unwin Ltd.
Jünger, E. (1926). *Der kampf als inneres erlebnis*. E. S. Mittler & Sohn.
Jünger, E. (1981). *Der arbeiter–herrschaft und gestalt*. Klett-Cotta.
Lévinas, E. (1988). Useless suffering. In R. Bernasconi & D. Wood (Eds.), *The provocation of lévinas* (pp. 156–167). Routledge.
Merali, Z. (2014). Stephen Hawking: "there are no black holes". *Nature*, 14583. https://doi.org/10.1038/nature.2014.14583
Minsky, M. (2007). *The emotion machine: commonsense thinking, artificial intelligence, and the future of the human mind*. Simon & Schuster.
Moravec, H. (1988). *Mind children: the future of robot and human intelligence*. Harvard University Press.
Nietzsche, F. W. (1980). *Nachgelassene fragmente 1885–1887, KSA* (Vol. 2). G. Colli-M. Montinari.
Paić, Ž. (2018). The permanent condition of war-and-peace: from the total mobilization to the absolute construction of the event. *International Journal of Philosophy, 6*(2), 40–54.
Paić, Ž. (2022). *Nemesis: aporias of the political and politics*. Naklada Breza.
Paić, Ž. (2023). *The superfluity of the human: reflections on the posthuman condition*. Schwabe Verlag.
Penrose, R., & Severino, E. (2022). In F. Scardigli (Ed.), *Artificial intelligence vs. natural intelligence*. Springer.
Scheler, M. (2016). *Krieg und aufbau*. Verlag der Weißen Bücher.
Schelling, F. W. J. (2006). *Philosophical investigations into the essence of human freedom*. SUNY Press.
Schmitt, C. (1997). *Der nomos der erde im völkerrecht des jus publicum Europaeum*. Duncker & Humblot.
Sloterdijk, P. (1987). *Kopernikanische mobilisierung und ptolemäische Abrüstung*. Suhrkamp.

Sloterdijk, P. (2006). *Zorn und zeit: politisch-psychologischer versuch*. Suhrkamp.
Sorgner, S. L. (2023). *We have always been cyborgs: digital data, gene technologies, and an ethics of transhumanism*. Bristol University Press.
Wiener, N. (2019). *Cybernetics or control and communication in the animal and the machine*. MIT Press.

CHAPTER 8

Freedom as an Event: Groundlessness as the Openness of the World

8.1 Introduction

Freedom as an event? Against the understanding of structural anthropology by Claude Lévy-Strauss, according to which all cultures are equal and have the same role in what is secularized as modern civilization, and within the framework of Hegel's metaphysics and speculative-dialectical thinking, the absolute has the even grander name of "world spirit", I will immediately place at the beginning the extremely "politically incorrect" axiom. Philosophy is seen as part of the West's historical mission, rooted in the primordial thought of the Greeks. Admittedly, Indian and Chinese philosophy undoubtedly function as a religious path of cosmic emptiness and as a pragmatic "wisdom" for maintaining the imperial order. Jewish and Arabic philosophy is onto-theology as a synthesis of eschatology and eternal justice. However, only the Greeks think of the historical existence of man from the most profound being of that which transcends any ontology as such and makes it possible at all. It is nothing but freedom as an event of openness to the meaning of Being. In Heidegger's lectures throughout his thought life, it was obvious that philosophy cannot be a mere logical thought, nor a worldview, just as it is not practical wisdom or an instruction for action, but thinking with its task or meaning and mission in the eminently non-metaphysical meaning of designing of life starting from the original freedom which is neither a Being nor a being,

Ž. Paić, *The Technoscientific Turn of Philosophy*,
https://doi.org/10.1007/978-3-032-21081-4_8

but simply an event of openness. In the book, *What Is Called Thinking?* Heidegger also said:

1. Thinking does not lead to any knowledge as sciences do.
2. Thinking does not impart any practical life wisdom.
3. Thinking does not solve the puzzle of the world.
4. Thinking does not have any immediate power to act. (Heidegger, 2002, p. 163).

Suppose thinking has as its "goal" merely thinking. In that case, philosophy should be the origin and historical destiny of this thinking, as it transforms Heraclitus and Parmenides to the end of metaphysics in cybernetics. The emergence of the technosphere , with artificial intelligence and artificial intuition, must remain extremely eccentric and experimental, an event of freedom. Philosophy does not regard freedom as an external or an internal object of reflection. This would mean that freedom exists as an "object" somewhere in the world, which can be established only as a principle of subjective thinking, as in the metaphysics of the modern Era, especially as presented by Kant, which posits the world in a space and time of infinite progress.

After all, even Hegel, who clearly shows that the essence of philosophy denotes the event of the synthesis of mind and reality from the impulse of absolute freedom of thought, defines history as a theodicy of the world spirit in the process of moving from the lowest to the highest stage of its realization. In the first part of the *Philosophy of History,* Hegel provides a detailed discussion of the Oriental world, encompassing China, India, Persia, Assyria, Babylonia, Syria, Judea, and Egypt. The second part, of course, is dedicated to the Greek world, and the third to Roman and medieval Christianity. Thus, the fourth part of the rise of the world spirit to absolute self-awareness would reach its peak in the Germanic Empire, culminating in the ideas of revolution and enlightenment. In practice, the modern principle of freedom originated in France. Still, its philosophical explanation is rooted in German thought, as evidenced by Kant's concepts of subjectivity and the autonomy of freedom. It is, therefore, the history of the world as the metaphysical history of the West.[1]

[1] It is, therefore, the history of the world as the metaphysical history of the West.

Freedom is immanent in philosophy because it is its fundamental "immovable mover", to use Aristotle's concept of God. Although its definition appears in various discursive explanations from Spinoza to Hegel and Schelling, and from Marx to Heidegger, it is evident that freedom is tautological: that which does not require any external power or control over action for its existence. Self-causation leads to freedom to self-purpose (Sutlić, 2022, pp. 145–146). This applies equally to the possibility of self-leadership and self-management of the posthuman set of thoughts and actions of the technosphere, understood as *autopoiesis*. Freedom is not only a question of the autonomy of the mind and the freedom of the human, but also a question of the correlation between *logos* and *psyche*. Therefore, man cannot consider any other being in the natural environment, such as an animal, in a Cartesian way as a mere object or *an actual extension*. It seems that this cosmic-natural-human-posthuman irreducibility penetrates the abyss of the groundlessness of freedom and opens it to the possibilities of creating a new world.

That's why, to understand the essence of freedom, it's necessary to exclude any mediating or enlightening reasoning about awareness as a process of realizing that freedom always means self-awareness of the inherent limits of one's being and its position in the world. Perhaps this was best outlined by Gilles Deleuze, who understood Greek philosophy as a contingent event of freedom, not a historical necessity, and that in the synthesis *of the polis* and the performance of life itself, by which only Greeks have the possibility for foreigners and non-Greeks to become citizens and to share the common destiny responsibility for the future of freedom under the auspices of democracy. Therefore, for Deleuze, the philosophical concept of immanence is always political, as it constitutes the condition of possibility for thinking from the eccentric essence of freedom (Deleuze & Guattari, 1994).

Why, then, is philosophy the only one of all forms of spiritual Being that enables history to become world history and carries within itself the sign of universality in the tendency against this or that pernicious particularism, whether imbued with the military mystique of Religion-Nation-Culture or cloaked in new clothing or some pseudo-cosmopolitanism? Why not myth, religion, art, and science, but only philosophy, which with Hegel, the today so denounced and almost "neutralized" and "suspended" thinker of the realization of Western metaphysics, reached the peak of its "mission", "task", and power to establish reality itself as mind-based principles of leading life in the form of an absolute science of freedom, not a science of liberty or any applied "philosophy of freedom"?

Today, due to the so-called deconstruction of Western history as logocentrism and political colonialism derived from it, the suppression of the Other and its irreducibility, we are no longer inclined to "mystify" the power of philosophy in its construction of the world and worlds (Derrida, 1967). I agree. But my consent is for entirely different reasons than from agreeing to this unconditional ethical-political turn by which the so-called Western civilization is freeing today mainly on a mere discursive-rhetorical level from its historically established black holes and dark shadows of the past that compromised even the idea of enlightenment as well as the modern project of the world-historical process of freeing the world from the horrors of religiopolitical dictatorships, tyranny, autocratic orders, and finally the monstrous totalitarianism of the twentieth century (Paić, 2022).

The philosophical notion of history presupposes a developed principle of freedom. What does this mean other than self-awareness as autonomy in what Kant defines as a moral act of unconditionality of action, or as what Hegel takes as the beginning of any further consideration of the absolute in the form of an idea that finally overcomes the so-called necessity of natural determination, according to which man is always in a state of dialectical relationship between slave and master. The developed principle of freedom belongs to the speculative level with which modern thinking begins, seeking to "free" man from external slavish purposes and to give him complete dignity, starting from his spiritual focus on what is irreducible to utility, pragmatics, and instrumentality. That is why, after the end, Hegel's absolute becomes the task and matter of philosophy in the thinking of the realization of freedom as (1) Marx's practice of work as scientific history, (2) Nietzsche's will to power as the eternal recurrence of the equal, and (3) Kierkegaard's existence as a thrust of singular faith in the divine source freedom as such in the abandoned and alienated world of aesthetic-ethical dead ends.

The entire post-Hegelian situation of the nineteenth century, marked by the discovery of non-alienated work, the will to power, and the religious experience of a man who communicates directly with God, is shaped by the effort to conceive of freedom as an irreducible event of thought that transcends the metaphysical boundaries of Western history thus far (Paić, 2021, pp. 63–116). We must by no means lightly pass over this triad of fundamental concepts of the end of metaphysics for the simple reason that work as life in the structure of the will to power as the unconditional rule of human existence in the form of an inhuman machine destroys the elementary freedom of Human as a whole being *logos* and *psyche* and transforms it into what belongs to the third order of cybernetics as the nihilism of the rule of artificial intelligence over man and his world. The problem with the so-called philosophical understanding of freedom in a world without dedication to the Other and concern for the world is that it becomes the fundamental question of the inauthenticity of man and the philosophy that provides him with such a limited justification.

The paradox is that every return of the so-called human world in modern times, in this contingency, chaos, and technoscientific models of the construction of autonomous objects rule, missing the essence of freedom because it does not perceive that freedom does not have its own "essence" and cannot be ontologically fixed on anything other than the empty place of its groundlessness. Heidegger gives the most significant "definition" of being human when, in 1929, in the treatise "The Essence of Reason", he claimed that freedom means the abyss of nothingness (Heidegger, 1978a, 1978b, pp. 123–173). This is the "uncanny", the most magnificent, and the most tragic, namely, that man exists without his essence, which constantly opens to him as a necessity of determination, responsibility, and openness of the dignity of Being itself, and which he must assume to justify its baseless existence.

Without freedom, history would be a blind necessity of nature incorporated in man, which would not allow him to realize himself as a free individual, as a representative of his own salvation and tragic fate, but as a slave subject of some despotic-tyrannical system of "history", which is simply the negation of the meaning of philosophy and of her complete determination and eccentricity as the opposite of what Schopenhauer and Nietzsche call will in the ontological sense of the synthesis of thought and action. As a fundamental driver of events, freedom must manifest itself in its three "ontological" spatiotemporal modes of appearance: chance and necessity, fact and contingency, and chaos and emergence.

The first opens philosophy as a possibility of thinking beyond reduction to myth, religion, art, and science, thereby establishing its autonomy in the age of the technosphere. The second is the one that has its political meaning of action, determination, and the creation of something new in history, starting from the idea of the sovereignty of the people and ending in the post-imperial sovereignty of an large/wide space *(Gro§raum)*, as Carl Schmitt defines the end of the epoch of the nation-state in the history of the world and the transition to the form of imperial order of power and their permanent struggle for rule over the territory of the Other. The third is the last sign of resistance to all forms of enslavement and the reduction of man to a thing-object-information within the system of rule of the posthuman condition, with *homo kybernetes* as the previous form of the "spiritual Being" of man in general.

8.2 Philosophy as a History of Freedom

Hannah Arendt, in the book *The Life of the Mind—The Groundbreaking Investigation on How We Think*, in the third part, "What Makes Us Think?" ponders the question of the origin of thought, the driving force that leads to philosophy as a question of the meaning of battle. In doing so, she engages in the sources of Heraclitus and Parmenides, as her thought teacher, Martin Heidegger, did throughout her life, particularly in lectures. In dealing with Heraclitus, we find the opinion presented.

> With the awareness of the invisible harmonious order *of the cosmos*, which manifests itself like other visible phenomena, although these have become obvious, the philosopher is amazed at the 'invisible harmony', which, according to Heraclitus, 'is better than the visible' (*harmony aphanes phaneres kreitton)*. (Arendt, 1978, p. 143)

In the following, Arendt demonstrates that in Heraclitus, this harmony is related to two fundamental concepts of early pre-Socratic thought: *physis* and *logos*. The first word means the same thing as being (*einai*) for Parmenides. At the same time, the second is a metaphysical synonym for thinking, which is not mere reason but the gifted ability to distinguish what belongs to good and evil. But harmony connects *physis* and *logos*: nature as being in its openness and thinking as insight into what is. Wondering at the fact that being and non-being are not, that is, being and non-being are not, is not possible without a simultaneous wonder at wonders, that there is harmony between being and thinking in the original understanding. The invisible is better than the visible because it is the condition of the possibility of the visible. The world's appearance must be based on what makes appearance possible in the first place. However,

Heraclitus is not Plato and an "idealist", just as he is not a "materialist" like Democritus. What Arendt derives in the development of this concept of wonder, with which philosophy begins, refers, therefore, to the question of where this invisible harmony comes from, that agreement and consonance, that harmonious action which, in the concept of physis and logos, gives the original thought the necessity of synthesis and the possibility of separation.

Heraclitus thinks of the idea of God in his own way. However, he does not express it here, except in the signification of *the logos* when he talks about Apollo, who in human language always indicates and shows the truth about Being, but not in the form of rational proof. Therefore, logos is always the one who collects and discerns the invisible weaves of harmony in the world, but cannot be equated with what makes this harmony possible. God has no logos but thinking; even in Aristotle's *Protreptikos*, it is closest in intention to the divine. The question of God, in Arendt's account, becomes the fundamental question of metaphysics, from Leibniz to Schelling. In 1929, Heidegger reinvigorated it in a profound way in his discussion, "What is metaphysics?" ("Was ist Metaphysik?"). When he said that the question of why there is a being at all, and not Nothing, becomes the decisive question of contemporary thinking in general, and as we can see, it is at the very origin of Western philosophy (Heidegger, 1978a, 1978b, pp. 103–121).

Arendt questions Spinoza's *causa sui* as a modern-day equivalent of Aristotle's first mover. It is about the so-called philosopher's god, who is claimed to be an unconditional necessity from which the freedom of the will can only arise, from which thought thinks all that is and what is not. But the question remains: where does the reason for the existence of this unconditional necessity, that is, the self-affirmation of the cause of itself, come from if the idea of God governs this world as an invisible harmony? Unlike Arendt, my understanding of being a god-philosopher is as follows. The harmony of Being and opinion forms the structure of the entire history of metaphysics from Heraclitus and Parmenides to Hegel and even Whitehead, with which he began the philosophical consideration of the concept of event in his fundamental book of procedural ontology *Process and Reality* in 1929, cannot be significantly changed even if, instead of the permanence of Being, we claim that anti-essentialist "there is" only becoming (*devenir*), as is the case in Deleuze's *Difference and Repetition*. God is therefore either suspended and neutralized, as in Hume and the empiricists, or materialistically negated, which is considered utterly unnecessary for the existence of the world and possible worlds.

After all, astrophysicist Stephen Hawking had two theories about black holes in the universe: an early and a late one. In the early days, Deleuze argued that a god was unnecessary to explain the Big Bang model of the universe's creation. This means, ontologically and without starting from Einstein's theory of relativity or quantum theory, that causa sui, or unconditional necessity, arises from the provision of the metaphysically understood first cause of movement, leading to movement itself and the dynamic structure of matter and energy. Hence, we could say by analogy that it is Spinozist pantheism without God. Towards the end of his life, Hawking rejected his postulatory atheism, as Max Scheler would say, and became a postulatory theist, because he was inclined to think that beyond the logic of creation, some spiritually or metaphysically initiated event is needed to happen in the "time" before the creation of the world. Is this the unconditional necessity of Heraclitus' invisible harmony that arises from the idea of God, and which decides before creation that it is better for a being to exist than not to exist?

Let's look at things this way. Instead of harmony, we can assume the world is chaotic, characterized by deviations from the rules. Lawfulness is the repetition of the same thing. Still, as Kierkegaard also knew, there are two types of repetition from which Deleuze partly took the notion of repetition as being or becoming something other and different. It partly derives from Nietzsche via Whitehead to Simondon. The first repetition is the Eleian, a Greek term denoting the past, and the second is modern and points to the future. The question that inevitably arises is whether this new, deterministic concept of chaos, with which, as is known, the butterfly from Hawaii causes the gushing geysers in Iceland with its wings, is just a different way of that unconditional necessity of happening or an ontological reason for ending any possible idea of the harmony of *physis* and *logos*, Being and opinions? Suppose chaos is a universal aspect of the universe. Does the concept of regularity and repetition stem from the repetition of the same, which is inherently non-harmonic and non-ordered?

Ultimately, does this mean the complete elimination of God from thinking about Being as such, in the sense that we can now claim, like Deleuze, that metaphysics instead of four has a new "trinity"—Being as becoming, difference and repetition in philosophy, art, and science—which is constantly transformed in dualities as in the notion of a film image that has its image-movement and image-time? The philosopher's God raises a question about the meaning of unconditional or contingent necessity; it is merely a modification of the dispute about necessity and freedom within the framework of Greek atomism, as seen in the debates between Democritus and Leucippus. The first advocated the necessity of atomic motion or determinism, which presupposes the existence of God. In contrast, the second advocated the declination of atoms or the free deviation from the established direction, thereby rendering the existence of God unnecessary.

Be that as it may, the principium rationis with which Leibniz arrives at the idea of a pre-stabilized harmony in the universe means seeing the unconditional necessity of God, which is reached through theodicy as a synthesis of mind and intuition, so thinking is necessarily twofold, rational and intuitive, sciences are first, and art secondly, it indicates that the facticity of the world derives from the idea of good. According to Leibniz, this is the best of all possible worlds, meaning that chaos would lead to its destruction. However, in the context of infinity, Leibniz preserves the idea of God's existence and creation; thus, creation originates from eternity, giving temporality, as Heraclitus would say, an invisible harmony. However, what if chaos is the condition for the possibility of any harmony, and if this harmony is the result of what Simondon calls a metastable balance in which the triad of information-matter-energy appears as a kind of Planck's constant?

In this way, the question about God did not become unnecessary; it was not eliminated, and it is not ontologically meaningless, because Kant explained it best by saying that it cannot be ontologically proven, which means neither scientifically proven nor disproven is the reason for his existence. Chaos is *prima materia* and appears as a transgression of metaphysics, as it is in the schizophrenia of the thinking of an Artaud. Metaphysics itself does not lose the reason for its madness, as Pessoa would say, by becoming the principle of monism and by the fact that, as in Spinoza, God is moved from transcendence to the realm of eternal immanence.

Everything that has been said here cannot be any solution to the Leibniz-Schelling question about why being is rather than Nothing, because it is not possible to do this in any other way than with the reasons of *logos*. God did not create the world out of his madness and irrationality, because then metaphysics would be pure schizophrenia of chaos as a principle. His creation would be illogical and irrational, and nothing could oppose him except madness, which, like Freud, does not want to become like a god at the end of the ballad of self-enlightenment, the perfect wisdom of the world, and the perfect good, if God is anything other than the metaphysical stability of the principle of sufficient reason. What sound logic is to us if the world is created by madness and chaos until the end of the world?

The problem of the unconditional necessity of God as the *causa sui* of thinking lies in the very act of thinking, which derives from its freedom as the necessity of contingency. We cannot metaphysically think of God nor the world, neither being nor thinking, if we do not introduce the concept that precedes it and which, in its freedom and liberation, enables both God as *causa sui* and thinking as a case of absolute contingency. In the twentieth century, Whitehead, Heidegger, and Deleuze developed this concept within the context of ontological considerations. It is about an event. In theology, it is a key concept concerning God's openness to existence and mission in the world. For philosophy, it is a key concept that emphasizes the openness of the meaning of battle and the creation of the world, stemming from a spiritual need for thinking that transcends external limitations, such as those imposed by theology, science, and art. This is a key concept in science because theoretical physics addresses the event horizon at the edge of black holes in the universe. After all, art denotes more than a concept, since it should be the creation of something new, and creativity produces events as pure contingencies.

Are we saved even in this terrible hour when the ominous prophets of the rule of the technosphere in the form of artificial intelligence announce the emergence of an overpowering new "self-awareness" that tends to have the possibility of being causa sui like God and in the name of good if it believes in it almost militantly like HAL 9000 from Kubrick's *2001: A Space Odyssey*, decided to turn us into nothing because it was her absolute will? Does this save thinking, provided that one of its symbolic forms, as Ernst Cassirer would say, does not oppress and negate the other? Let's see what we got instead of the new "trinity". The old metaphysical fourth or Heidegger's post-metaphysical *Geviert* of earth and sky, mortal and immortal?

We have received prospects of thinking for the future, and that no longer of theology as a mystery of faith, but as a possibility of change in the very essence of the philosopher god, without whom today the so-called science of God has no chance of surviving the unconditional necessity of technosphere rule. We have been given the last possibility of salvation from falling into the uncanny abyss of technoscientific nihilism, which no longer needs a superfluous man even for its function.

8.3 Freedom Within the Borders of *Gro§raum*

The twentieth century has been marked by the confrontation between liberal democracy and totalitarian political orders worldwide. In Hayek's words, the Cold War was fought ideologically between the ideas of "freedom" and "slavery". Of course, this Manichaeism is a construction tailored to the needs of a linear dialectic of history, which was, after all, the one performed by Fukuyama in 1992 in his book, *The End of History* and the Last Man, in the footsteps of Kojève and his analysis of Hegel's *Phenomenology of Spirit* (Fukuyama, 2006). Is the twenty-first century, and we are in its third decade, a continuation of the same binary logic by other means, or perhaps something else? My answer is explicitly stated in the book *The Return of Totalitarianism—Ideology, Terror, and Total Control*. In the last chapter, which is entitled "Power, Mass, and Brutality: Putin's War Machine" (Paić, 2022, pp. 197–218), I tried to show that the nature of the so-called autocratic rule of Putin's war machine of a neo-imperial character, which is neither the restoration of Nazism nor Stalinism by other means, except on the symbolic-imaginary side of the patchwork of ideology as terror and total control outwardly and inwardly, i.e., by colonizing the so-called vassal states and state repression of political opponents in the formal democratic functioning of the state and society. Instead, it is a new political phenomenon: the construction of an authoritarian-autocratic order that, admittedly, bears some analogical resemblance to totalitarianism. However, what distinguishes it is the triad of a new *ideology, terror, and total control.*

What does that mean? Nothing else but that we encounter an "empty centre of power", so an autocratic Leader, unlike a totalitarian dictator, in this case, Putin or the Chinese ruler Xi Jinping, no longer rules by requiring the collective mind and organization of the Communist Party as a secularized deity over the country. The party has been replaced by a national-ideological system of terror and total control of the state as a new empire that constructs the subject of government, i.e., "the people", either populist or fundamentalist. In Putin's case, it is a neo-traditional alliance of icon and axe, Russian nationalism and Orthodoxy, and in the Chinese case, an effort to show the evolution of the idea of the great

Chinese empire from the period of ancient dynasties through the communist nation-building with Mao Zedong to the present day as a continuity of history in an eschatological sense. Autocratic orders are a triad of hybridity, eclecticism, and syncretism. What we would call fascism 2.0 and Stalinism 2.0 take place in them, but no longer in such a way that, like Marx, we could say that history repeats itself the first time as a tragedy and the second time as a farce. What makes defining this "new totalitarianism" of the twenty-first century challenging is its political "essence".

Namely, all the authoritative classical theorists of totalitarianism, from Hanna Arendt, Franz Neumann, Carl J. Friedrich, and Raymond Aron claim that autocracy in the way of governing the state and society is beyond the liberal-democratic principles of the modern age such as the separation of state and society, state and religion, tripartite legislative, executive and judicial authorities, which establishes the logic of Power with the help of the Leader-Party-Army triad, i.e., the repressive enforcement apparatus. However, the problem stems from the fact that totalitarianism in the twentieth century was a sign of the total subjugation of the Other by its destruction and extermination in concentration camps. The Nazi camp system for the deportation of Jews and the Stalinist Gulag for the deportation of "internal enemies" was based on the idea of permanent terror and the destruction of all forms of liberal relations in what Hegel calls the structure of the objective spirit from the family, society, to the state. The twenty-first century is no longer the century of concentration camps for the condemned Others but the total control of the Other with the help of the technosphere, a combination of artificial intelligence and biopolitics in the Agambenian sense of the word, which means that instead of exterminating the Other, it is neutralizing and suspending it. The persecution and expulsion of minorities, ethnic groups, and political opponents are carried out as a strategy for cleaning up the geopolitical space with subtle methods, in Deleuzian terms, of deterritorialization/reterritorialization.

To further clarify this essential difference, which, as Freud would say, is in the spirit of the "narcissism of minor differences", we will return to the analysis of political authority in Hannah Arendt. Namely, she says that everyone.

> Behind the liberal identification of totalitarianism with authoritarianism, and the concomitant inclination to see "totalitarian" trends in every authoritarian limitation of freedom, lies an older confusion of authority

> with tyranny, and of legitimate power with violence. The difference between tyranny and authoritarian government has always been that the tyrant rules in accordance with his own will and interest, whereas even the most draconic authoritarian government is bound by laws. (Arendt, 1961, p. 97)

Arendt distinguishes between tyrannical, authoritarian, and totalitarian systems. However, what determines all three historically epochal established political orders of government, from the ancient Greek tyrannies to Nazism and Communism in the twentieth century, and to Putin and Jinping, is nothing but the self-rule of rulers with the unconditional prerogatives of Power. Autocracy, therefore, is the essence of rule with the help of the unlimited use of violence and various techniques of subjugation and colonization of the Other. It is based on authority, but not as the obedience of subjects to the Leader transmitted by tradition, but as a kind of transcendent form of Power that presupposes a vertical structure of its execution. An absolute break with the Enlightenment eschatology of Mind and the cosmopolitan order of the rule of knowledge and rationality marked the twentieth century.

In addition, totalitarianism was a form of political authority that we call singular and irreducible because it had no historical precedent. That is why many theoreticians, at all costs, tried to find any connections, even occult and mysterious, with tradition so that the logic of causality in the scientific sense could help explain *Auschwitz* and *Kolyma*. Arendt put the matter most radically by defining totalitarian rule in terms of contingency, emergence, and the transcending of the Enlightenment's vision of the future. The key to the rise of totalitarianism lies in the disintegration of the modern concept of the nation-state's sovereignty and the transition to a state best defined by Carl Schmitt in his work Nomos of the Earth, in which he employs the powerful German term *Großraum* (Schmitt, 1974).

The space that extends beyond the borders of land and sea in the modern order. The sovereignty of nations becomes the agent that destroys the metaphysical notion of world history. Now, namely, it is a question of a global-planetary order of total mobilization of technoscience, militarism, and the process by which informational-cognitive capitalism no longer conquers the "desolate land" and enslaves its inhabitants as the original peoples in the system of colonialism, but creates conditions for neutralizing and suspending the freedom of the Other through the process that Deleuze and Guattari call the axiomatic of deterritorialization/reterritorialization, that is, displacement and settlement in the logic of devastation of the land and its transformation into surplus value and profit for future investment (Deleuze & Guattari, 1987).

Why is autocracy the essence of totalitarianism, and at the same time, what does not necessarily reach its form and practice of realizing radical evil in history? Because it is about the mode by which Power is established as a political state of rule over the Other without the external legitimacy of the state and society. Autocratic leaders are tyrants and dictators without tyranny and dictatorship. In the twenty-first century, their framework of action is the hybridity and fluidity of the network, by which three fundamental ideologies, as political articulations of ideas, govern the *Großraum*. These are fundamentalism, populism, and neoliberalism. Let us examine how Putin manages Russia's economy and compare it with China's. Of course, capitalist industrialization and technologization are carried out with the help of oligarchies and state-party managers. This is why the ownership structure is corporate imperial, as exemplified by the Chinese giant HUAWEI. What gives autocratic rule in the twenty-first century a comparative advantage over all forms of traditionally "weak" liberal democracies is that Arendt denies the paramount importance of authority. It is about normalizing violence and all kinds of repression for the Leader's pragmatic interests and his vision of ruling the world.

In other words, hybrid autocratic orders in the twenty-first century are networked with each other and find themselves, paradoxically, in a constant "conflict dialogue" and consensus with liberal democracies because corporate capitalism is anything but the apotheosis of tolerance and trust. On the contrary, corporate capitalism is based on a Leader who is a narcissistic-pathological autocrat, such as Donald Trump. Therefore, the essence of autocratic rule today, within the triad of ideology, terror, and total control, presupposes the deconstruction of the traditional concept of authority. It is no longer any external or internal power to which the Others unconditionally and reasonably obey, but rather the Power of brachial rule, aided by structural violence, that unites economy, politics, and culture as autopoietic systems networked in *Großraum*.

The essence of autocracy designates the unconditional will to power within the limits of total control, which cannot be overthrown by the Leader's death or violent overthrow, because the order becomes, as Gilbert Simondon would say, in a state of metastable equilibrium (Simondon, 1994). The mysterious power of tradition guarantees it.

With the help of political religion as the spiritual substance of government, it continues with the alliance of "icon and axe" even when the Leader is neutralized and suspended because the essence of autocracy in the twenty-first century appears as the establishment of the logic of the state of emergency *(Ausnahmezustand)* on a global-planetary level. This also applies to democracies and dictatorships. Is there any way out of this *no-way-out* situation? Has freedom in its monstrous power of disobedience remained without its substance and subject, traditionally metaphysically speaking, and that means without the sovereignty of the people in the logic *of Gro§raum* and the loss of the existential foothold of man himself who falls into the abyss of terrifying indifference to the universal freedom of the cosmopolitan order without which philosophy as an eccentric event of thought in the openness of the world does not have its final justification? The fear of the possibility of the emergence of a "new totalitarianism" brings freedom to the extreme limits of human dehumanization, primarily because man is no longer the goal and purpose of the developed system of autopoietic action of the technosphere. It is its structure, function, and assembly of the pragmatics of knowledge as know-how by which the visualization of life makes language, the telling essence of thought, inoperable and obsolete for what is becoming a question of the current future.

8.4 Technosphere and Autopoiesis—The End of Freedom?

One of Nietzsche's most radically written late works, *A Genealogy of Morals*, was composed between July 10 and the end of August 1887 in his summer residence of Sils-Maria in the Swiss Alps. Morality is an essential aspect of metaphysics, rooted in the ideas of Plato, Judaism, and Christianity, which have shaped the ascetic ideal. Instead of an apology for the abundance and vitality of life that adorned the Greek world of pre-Socratic philosophy and art, everything is moving towards denying life, renouncing the body and its sublime powers. Suppose the term corresponding to Nietzsche's artistic aesthetics is precisely physiology. In that case, morality as a set of values is a necessary ascetic way of life aimed at overcoming suffering. What follows from this is that moral judgement cannot always be particular or universal. Morality represents the decadence of life. When asceticism in the form of Christianity is embodied, suffering rises self-consciously to the fundamental principle of leading life.

We are faced with the question that Nietzsche poses dramatically in the 28th fragment of this writing. This question and Nietzsche's answer, as presented in my book *The Superfluity of the Human—Reflections on the Posthuman Condition*, offer a path to immortality without a secret (Paić, 2023, pp. 87–108). Nietzsche, therefore, says.

> Man, the bravest animal and the one most inured to suffering, does *not* repudiate suffering in itself: he *wills* it, he even seeks it out, provided that he is shown a meaning for it, a *purpose* of suffering. *Not* suffering, but the senselessness of suffering was the curse which till then lay spread over humanity, *and the ascetic ideal gave it a meaning!* It was up till then the only meaning, but any meaning is better than no meaning; the ascetic ideal was in that connection the *"faute de mieux" par excellence* that existed at that time. In that, ideal suffering *found an explanation*; the tremendous gap seemed filled; the door to all suicidal Nihilism was closed. (Nietzsche, 2013, p. 210)

Suffering presupposes illness of the body and, at the same time, sacrifice as a pledge of redemption. Therefore, the meaning of suffering can only be metaphysically and religiously determined by the expectation of eternal bliss in another life. When Nietzsche asks the question "Why man?", the animal within the human must be overcome morally, as Kant postulated in the categorical imperative. However, the central question in the *Genealogy of Morals* concerns the genealogy of the *posthuman condition*. This means that only from the "ontological" position of the superhuman can we leave the biological one, as predestination and the necessity of enduring the suffering that affects human destiny. Instead, what transcends the limits of any "ascetic ideal" is nothing human-too-human but transhuman. That is why a radical critique of Platonism and its inherent Judeo-Christian morality is essential to overcoming resentment. It is no longer a praise of a tragic life like the Greeks, but at the very least, an advocacy of empty messianism, driven by the so-called hope for the future.

The new morality removes the "foundation" for the so-called higher goals. Suffering ceases to be an alibi for ascetics and religious exegetes when there is no longer a reason for messianism. Nietzsche's key argument against the "slave psychology of Christianity" is that without suffering and sacrifice of the body, redemption does not have its ultimate purpose. But what is most monstrous in all this arises as the nihilism of the end of metaphysics when man realizes that he has only the will for nothing left because everything else has become superfluous. The redundant man is not the last man without morals, but the first *homo kybernetes* whose essence is determined by what is no longer fate or destiny. Absolute freedom gives the power to transcend the history of inhumanity.

Chaosmos denotes a key notion not only of the late Deleuze but also of contemporary philosophy and cosmology. Although Deleuze himself appropriated it and gave it a new meaning by drawing on Joyce, Gombrowicz, and Borges, its actual "ontological" origin lies in the Stoics. The proper place of permeation of thought and Being lies under the auspices of the cybernetic theory of complexity and emergence (Deleuze, 1969). I do not intend to delve into the genealogy of a hybrid assembly, which directly addresses the core of modernity. It is enough to point out that at the end of the first and in more detail in the second volume of my *Technosphere*, as part of the analysis of the concepts of information, feedback, control, and communication in Norbert Wiener, Godhard Günther, and Max Bense and the idea of metastable equilibrium by Gilbert Simondon, this way of thinking of trans-classical logic develops either or, simultaneity and correlation of space and time as virtual actualization (Paić, 2018).

Everything that adorned the postmodern aesthetics of Alexandrinism, from Eco to Borges, with the triad of concepts syncretism, eclecticism, and hybridity, now appears as the characteristics of a non-dialectical synthesis. What does this mean other than that what belongs to the technosphere determines the nature of the new limits of appearance? Research in artificial intelligence provides strong support for this assumption. After all, cybernetics, in its essence, is not self-leadership, self-management, and self-organization based on the logic of nature as such, in the sense of homeostasis, but rather the subjugation of nature to human and non-human purposes of technical existence. Being a machine marks a turning point in understanding the relationship between Being and thinking, as a new technology for the production and distribution of information. Therefore, the *cosmos cannot be a synthesis of the original disorder and order at a higher level of speculation and reflection, but rather a cybernetic permeation of the original and the produced, starting with* the reign of the "Big Third".

The ancient peoples envisioned the cosmos, nature, and humanity as the result of a theomachy, from which the ever-supreme, mighty God, who rules the world as a union of heaven and earth, based on the established order in all things, from mysticism to human nature, from cosmology to politics and art, emerges as the victor. Harmony and balance have been attributed to ontological features in Western metaphysics from the Greeks to quantum theory. Everything has its fixed place in space and time, and Being is constant, eternal, and unchanging. The term chaos in Hesiod's *Theogony* signifies that which is formless and monstrously incongruous in its irreducibility. Order, as taxis and cosmos denote, marks the emergence of balance between the relationship of gods and humans. That is why the term with the same meaning for Parmenides as being is nature—*physis*. Duality remains necessary. The only question is how they are established, from above or below. If the first case is valid, we discuss transcendence as the principle of shaping the apparent world. Suppose immanence is the one that leads the game. In that case, we are in the univocal Being, in a single world of empiricism for which, admittedly, as in Spinoza, God can also be a substance as a recognized necessity.

But both the transcendental and what belongs to the immanentism of the world, instead of the Manichaean struggle for life and death in epistemology, are now interwoven in something "third". Kant and Hume become authoritative in Deleuze's "transcendental empiricism" thinking. What was utterly separated is now synthesized, but the dualities remain only in a different metaphysical relationship. What designates the term *chaosmos?*—chaos as order or order as chaos? Both. But this is, or the correlative simultaneity of what is happening in the world, both good and bad, both authentic and vulgar, both democracy and dictatorship, both progress and decline and man and intelligent machine, cannot be feasible without that "Great Third", *tertium datur* which cybernetically directs the trajectory of the wandering irreversibility of "thinking" and "Being". *Chaosmos* cannot be a cybernetic "God" of the technosphere because the new quaternity of causes, in contrast to Aristotle's classical scheme of causality and teleology, becomes the one that techno-genetically and auto-poetically creates new events, rather than copying and imitating what already is, as Being-God-World-Human.

Chaosmos denotes an event produced by thought, realized as cybernetics within the technosphere. Instead of the harmony and balance of nature, everything accelerates infinitely, becomes different and different from the same, loses the substantial determination of the One, and collapses into a technical abyss. Nothingness can no longer be negation and nothingness, but nothing created from thought as an event of contingency no longer has necessity. Freedom precedes *chaosmos*, and its irreducibility is *Unheimlich* because it applies to animals, humans, and creatures created by artificial intelligence. Freedom outside the anthropo-humanistic circle of metaphysics becomes genuinely "chaosmotic", for which our opinion can no longer be authoritative.

When metaphysics, with its monumental, grandiose, and illusory framework of order, harmony, and synchronicity, no longer works, it seems evident that we are at the end of history, as the end of metaphysics in the age of the technosphere, marked by nihilism, which functions as a metastable balance between chaos and new order. And that is why we are so traditionally faithful to any order from good old metaphysical history. However, let us forget about those illusions without coverage. When thinking of "thinks" as an event, everything is open in its indeterminacy of freedom, which is a sign of groundlessness as the openness of the world. Even if by some "miracle" we become resistant to this frantic transformation of conditions and processes and begin to think anew on the ancient traces of that "desolate land" through which we pass like sleepwalkers in an even deeper night than this global-planetary eclipse of the world, revealing the signs of a new dawn.

8.5 Conclusion

Human-as-experiment becomes a genuinely new "ontological experience" beyond all previous metaphysics because it has become the highest possibility of one's realization of freedom. This event, however, is not a matter of primordial tradition in which Greek philosophy takes place. Only with the help of what enables the emergence of the non-human in the scientific-technical process of liberating nature and the human in the practice of work or the will to power does man come to the awareness of the openness of history at the end of its metaphysical completion, not before. Thus, the uncanny experience of the experimentality of freedom is a historical event in which one gains an understanding of the meaning of Being as the freedom of the openness of the human and all other possible worlds, in nature and the cosmos, and the existential challenge of what comes from freedom itself as the essence of man. Isn't it significant that, among all the ideas in modern metaphysics—from Kant to Hegel—establishing the system and method of the speculative dialectic of history from ontology to ethics, politics, and aesthetics, freedom is the only one that lacks a foundation and the principle of sufficient reason? Heidegger, therefore, rightly called Schelling's book on freedom the most significant writing of classical German idealism (Heidegger, 1988).

Suppose freedom cannot be that which is causally deterministically set as a principle of action. In that case, one cannot think of freedom as liquid energy, but rather as the openness of Being, which enables and opens history. Metaphysics necessarily assumes freedom as the matrix of orientation towards the first cause and the last purpose. Therefore, the historical-epochal appearance of freedom in world history takes place through three paradigms of thinking and designing Being—ontologically through the work of established history with the help of philosophy as the source of science and technology, politically through entering the "great space" (*Großraum*) of post-imperial sovereignty in which the power of total mobilization of the technosphere and capital is established as the last border between autocratic and democratic orders, and finally cybernetic in which life itself becomes information-control-communication and freedom opens up as an existential appeal and response of man to his dehumanization. Freedom is neither something for Nothing, but precisely that historical openness to the possibility of an event which, in its irreducibility, denotes itself no longer thinkable philosophically, politically, or cybernetically. Such an event raises the question of the vitality of life itself in the era of universal nihilism, where groundlessness becomes a new openness of the world.

Note [1]In *the Philosophy of History*, Hegel makes the famous statement that "world history is this event of the development and real existence of the spirit...and this is a true *theodicy*, the justification of God in history". The only insight that can reconcile spirit with world history and reality is that what happens and will happen forever, not only is not without God, but his work as such is essential" (Hegel, 1970, p. 540).

References

Arendt, H. (1961). What is authority? In *Between past and future*. Viking Press.

Arendt, H. (1978). *The life of the mind–The groundbreaking investigation on how we think*. Harcourt, Verso.

Deleuze, G. & Guattari, F. (1994). *What is Philosophy?* (G. Burchell & H. Tomlinson, Trans.). Columbia University Press.

Deleuze, G. (1969). *Logique du sens*. Minuit.

Deleuze, G., & Guattari, F. (1987). *A thousand plateaus: Capitalism and Schizophrenia II*. Continuum.

Derrida, J. (1967). *De la grammatologie*. Minuit.

Fukuyama, F. (2006). *The end of history and the last man*. Free Press.
Hegel, G. W. F. (1970). *Vorlesungen über die Philosophie der Geschichte, Werke in zwanzig Bänden* (Vol. 12). Suhrkamp.
Heidegger, M. (1978a). Vom Wesen des Grundes. In *Wegmarken* (2nd expanded and revised ed.). V. Klostermann.
Heidegger, M. (1978b). Was ist Metaphysik? In *Wegmarken* (2nd expanded and revised ed., pp. S. 103–S. 121). V. Klostermann.
Heidegger, M. (1988). *Schelling: Vom Wesen der menschlichen Freiheit (1809)* (Vol. GA, Bd. 42). V. Klostermann.
Heidegger, M. (2002). *Was heisst Denken?* V. Klostermann.
Nietzsche, F. W. (2013). *The genealogy of morals: A polemic*. T.N. Foulis.
Paić, Ž. (2018). *"Black box of metaphysics": Cybernetics and the absolute time of the machine, Technosphere* (Vol. II). Sandorf and the Misanthrope.
Paić, Ž. (2021). *The Spheres of existence: Three studies on Kierkegaard*. Toronto University-Apocalypse.
Paić, Ž. (2022). *The return of totalitarianism: Ideology, terror, and total control*. Palgrave Macmillan.
Paić, Ž. (2023). *The superfluity of the human: Reflections on the posthuman conditions*. Schwabe Verlag.
Schmitt, C. (1974). *Der Nomos der Erde im Völkerrecht des Jus Publicum Europaeum* (2nd ed.). Duncker & Humblot.
Simondon, G. (1994). *Une pensée de l'individuation et de la technique*. Chatelet.
Sutlić, V. (2022). *Lectures on Nietzsche* (D. Barbarić, Ed.). Matica hrvatska.

CHAPTER 9

Conclusion

The technosphere that appears in the contemporary era of posthumanism, with the total and imperial power of reducing thinking to the rational and the poetic, or as Heidegger puts it in the lecture and writing *What Is Called Thinking? (Was hei§t Denken?)* determined by the difference between thinking as calculation (*Rechnen*) and thinking as reflection (*Dichten/Besinnung*), denotes an epoch-making feature of a radical change in the form of life (Heidegger, 2002) . With a technosphere, man and animal only "now" and "here" come to the truth about their relationship. Due to the significant threat of extinction to species and life on Earth, including both human and animal species, it may be possible to ask the question "what" it means to be an animal and "who" is a Human, outside the anthropocentric horizon. Of course, this question becomes essential because, at the end of history, the biological structure of what is called life, in all its forms, disappears in favour of the techno-structure and the techno-morphic creation of a new "life". The technosphere enables consideration of the relationship between the two. But no longer from the position of onto-theology (Being and God), but rather the place of techno-genesis of life (machine and aesthetics).

The singularity of the machine stems from the conditions that make it possible to combine the animate and the inanimate, nature and technology. The idea of the machine in the biocybernetics model of thinking, therefore, belongs to the synthesis of space and time as indeterminacy in

Ž. Paić, *The Technoscientific Turn of Philosophy*,
https://doi.org/10.1007/978-3-032-21081-4_9

the intervals of the creation of the "new". This synthesis, in turn, allows space to be compressed and time to be stretched to infinity. It might not be difficult to see that this becomes a feature of virtuality. Immersion *in* virtual space means being present in real time. The machine does not operate according to conjunction-disjunction logic. On the contrary, the machine embodies the concept of synthesizing "artificial life" (*A-life*) but does not show or represent "Being". Instead, the machine visualizes concepts as a computer would. It does this by programming the codes of completely autonomous communication between machines and people as objects. That is why the key idea of the digital age—the machine organization of all life in networked societies—has become interactive.

The technosphere might no longer be displayed in the common terms of the metaphysical tradition (*mimesis*). Hence, the collective power of creating a new aesthetic will now be constructed as a sublime experience of making things beyond the limits of nature and humans. The thing itself creates, and the things are not the same as things in themselves. The creature was now created through the technological reproduction of the Being. It is conceived, however, as a technical object other than a living object. But what gives it specificity in the anthropomorphic world of functions and structures is the system's applicability, which nowadays is represented as a techno-genetic phenomenon that emerges as a way of regulating social relations. An information circuit of communicative action was created, and the creation in terms of a technical object must be available for Human use. Both, however, form the cyberspace environment as a management system composed of artefacts and living things immersed in the network.

Let us not forget that the integration of technology into human life cannot be just a new frontier of post-biological evolution. If what I call the technosphere lies beyond any distinction between animate and inanimate, human and inhuman, then everything that belongs to artificial intelligence and its thinking as *calculation-planning-construction* becomes what is fundamentally different from Hegel's absolute, which produces the world as a projection of one's work of self-consciousness in the circle of metaphysical determinations. The technosphere, unlike divine rationality and intuition, which provide direct insight into what has been, is, and will be, becomes infinitely adaptable and plastic in response to any change in what Alfred N. Whitehead called an event in the history of cosmic epochs (Whitehead, 1978). Philosophically speaking, thinking of the essence of the technosphere can no longer be possible, nor does such an endeavour prove meaningful because it is done by the very logic of techno-genesis, by visualizing events as a virtual actualization of becoming. When information suspends the meaning of Being in its openness, we find ourselves in the lobby of a post- and transhumanist area with no signal. Nothing comes from outside or inside. Everything happens in simultaneous non-simultaneity, because here and now, as labels for the classical ontology of space and time, are no longer decisive.

When philosophy is reduced to technoscientific epistemology, or when its essence and task are reduced to bioethical discussions of the status of cyborgs, androids, and apparatus of consciousness in telematic societies of control today, then philosophy is left only enough to reflect on techno-futurism and techno-biologism by other means in contrast to scientists and cybernetic artists. Rather than speculation and reflection on Being as the objectivity of the object, this new philosophical thinking should focus on a synthetic reflection on the integrated circuits of contemporary technoscience experiments.

The key concepts of *singularity, technosphere,* and *autopoiesis* offer latent possibilities for creating an entirely new form of interaction between humans and non-humans. Aesthetic thinking, therefore, has the ontological-epistemological task of thinking and articulating the matrix with a new categorical system of fluid concepts as a different position of art beyond the service to superfluous Human. What is already coming becomes a sign of the rule of synesthetic transformation over all forms of existence, not only human subjectivity, but also evaporates into the ether like a bubble.

References

Heidegger, M. (2002). *Was heisst Denken?* (Vol. GA. 8). V. Klostermann.

Whitehead, A. N. (1978). *Process and reality: An essay in cosmology.* The Free Press. (Original work published 1929)

Index

A

C

Ž. Paić, *The Technoscientific Turn of Philosophy*,
https://doi.org/10.1007/978-3-032-21081-4

The manufacturer's authorised representative in the EU is Springer Nature Customer Service Centre GmbH, Europaplatz 3, 69115 Heidelberg, Germany. If you have any concerns regarding our products, please contact ProductSafety@springernature.com

Printed and bound by CPI Group (UK) Ltd, Croydon, CR0 4YY
22/09/2026
02220991-0002